Web Site Stats: Tracking Hits and Analyzing Traffic

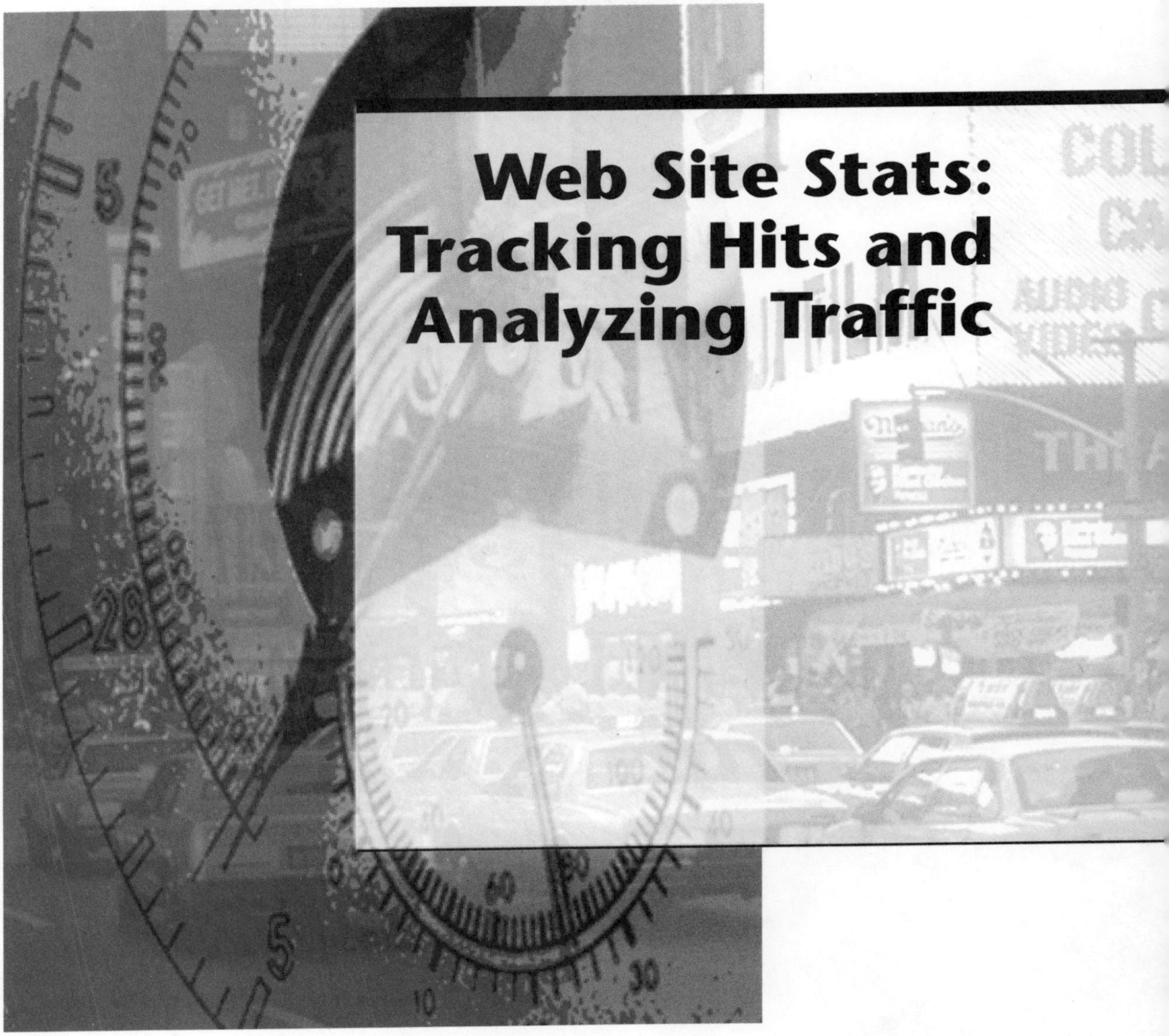

Web Site Stats: Tracking Hits and Analyzing Traffic

Rick Stout

Osborne **McGraw-Hill**

Berkeley New York St. Louis San Francisco
Auckland Bogotá Hamburg London Madrid
Mexico City Milan Montreal New Delhi Panama City
Paris São Paulo Singapore Sydney
Tokyo Toronto

Osborne/**McGraw-Hill**
2600 Tenth Street
Berkeley, California 94710
U.S.A.

For information on translations or book distributors outside the U.S.A., or to arrange bulk purchase discounts for sales promotions, premiums, or fundraisers, please contact Osborne/**McGraw-Hill** at the above address.

Web Site Stats: Tracking Hits and Analyzing Traffic

1234567890 DOC 9987

ISBN 0-07-882236-X

Publisher
Brandon A. Nordin

Acquisitions Editor
Megg Bonar

Project Editors
Janet Walden
Cynthia Douglas

Editorial Assistant
Gordon Hurd

Technical Editor
Morgan Davis

Copy Editor
Kathy Hashimoto

Proofreader
Pat Mannion

Indexer
David Heiret

Computer Designer
Richard Whitaker

Illustrator
Roberta Steele

Cover Design
Ted Mader Associates, Inc.

Quality Control Specialist
Joe Scuderi

About the Author

Rick Stout is an internationally recognized expert in the fields of computers and the Internet, and has authored and coauthored numerous books, including *The World Wide Web Complete Reference, The Internet Complete Reference, Peter Norton's Introduction to Computers*, and several volumes of the wildly successful Internet Yellow Pages series. He has also developed and maintains a collection of web sites ranging from commercial businesses to sites dedicated to his previous books.

Contents at a Glance

Part 1 The Gory Details

Part 2 Analyzing Server Log Data

Part 3 Log Analysis Tools and Services

Part 4 Marketing and Advertising

Part 5 Appendixes

Contents

Part 2

Analyzing Server Log Data

Part 4

Marketing and Advertising

PART 5

Appendixes

Acknowledgments

First off, I'd like to thank the best team in computer book publishing—my acquisitions editor Megg Bonar and my project editor Janet Walden. Despite my inability to meet even a single deadline on this book, Megg and Janet were always gracious and understanding. In fact, they're downright fun to work with. This was my first project with Megg and Janet, and I very much look forward to working with them again.

On all of the previous books I've done with Osborne/McGraw-Hill, Scott Rogers was my acquisitions editor. In the middle of my last book (the one prior to this—*The World Wide Web Complete Reference*), Scott was promoted to executive editor (and ultimately to editor-in-chief). This means that now he spends most of his time flying around the country making celebrity appearances at corporate meetings and book conventions. But I still get to talk to Scott on occasion. In fact, he offered the inspiration for this book, as well as some good ideas on the content, which I promptly used. For these, I am grateful.

I haven't worked with a great many publishing companies (only three, to be exact—but who's counting). But I doubt there's a better organization in the business than Osborne. The people are real and they work tirelessly and often without adequate recognition to produce and market the quality books they do. I wish I could thank everyone who has had a hand in producing this book, but that would be nearly everyone in the company. However, I do want to thank several people individually: Kathy Hashimoto, my copyeditor. Kathy did a great job, and I hope to work with her again. Cynthia Douglas, the associate project editor, handled the book from the page proof stage on, coordinating my corrections with those of the proofreader, Pat Mannion, and attending to all manner of last-minute details. And thanks to the folks in production for making the best of some awkward layout challenges that a few chapters in the book presented. Thanks also to David Heiret for doing a great job on the index.

My thanks also goes to Morgan Davis for doing the tech edit on this book. Morgan was my coauthor on *The Internet Science, Research & Technology Yellow Pages*, and is the Director of Operations for CTS Network Services, the finest Internet Service Provider in Southern California. He's a friend and all around good guy. In addition to doing an excellent tech edit and saving me several times, Morgan gave me an ISP's perspective when I needed one. Thanks Morgan.

I also thank my good friend Tom McPhail. Throughout this book, I used Tom's golf site (www.golfcircuit.com) as a frequent guinea pig. Tom also shared a great deal of his hard-earned experience at working with advertisers, ad agencies, and ad networks.

Finally, I would like to extend my sincere gratitude to each of the people and the companies that contributed to the CD at the back of this book. Pulling together this CD was a grueling and time-consuming experience. But getting to know and working with each of these people proved to be a rewarding and enriching experience.

Eli Shapera	e.g. Software
Brent Haliverton	Group Cortex
Midori Chan Jessamy Cook-Tate	Intersé
Patrick Benard Matt Cutler	net.Genesis
John Keyes Peter Sidewater	Open Market
Dominic Tassone	Streams
Erin Gaffeny Jean Hagen Michael Tchong	I/PRO
Yvette Herrera	NetCount, LLC
Charles Smith	Real Media
Jeff Spillers	WebThreads

Introduction

Now that millions of people and companies are on the web with their own home pages and web sites and millions more are browsing the web and reading them, the time has come to ask ourselves if it's all really worth it. For years, we've been hearing about the commercial value of the Internet, and that any company worth its salt would have its own web site. Numerous companies and organizations have taken the plunge and developed web sites simply to have presence on the web, to sell products or services, or to sell advertising space.

Now it's time to reconcile the books and find out how many people are actually visiting our web sites, who they are, what they want, and what they do at our sites while they're there. Until relatively recently, there weren't too many options. You could use one of a handful of programs freely available on the Net. But most of these programs were (are) unsupported or undersupported, and offer only basic statistics of dubious benefit.

With this book, you will learn what information web servers capture about readers, what it's important to measure, and how to get the most out of your server log files. You will also get a rundown on the top software packages and service companies doing log analysis. We will pit the products and services head to head to give you the vital information you need to decide what is right for you. Finally, you will get a great introduction to advertising on the web learning about the mechanics of banner ads, ad servers, and ad networks.

How to Use this Book

Of course, the best course to take is to read the book from cover to cover. But, no doubt, not every chapter will be particularly interesting to everyone. If you're completely new to this subject and you know you will need to get into the technical details of traffic analysis, you'll want to start at the very beginning with Chapter 1. In Chapter 1 (and in the rest of Part 1), I explain the nuts and bolts of web server log files. If you know these files, you know everything that it is possible for a web server to record in a log file, and thus what it's possible to know about the visitors to your web site.

If, or when, you are generally familiar with the server log files, move into Part 2. In Part 2, we will grab a collection of hits out of an actual production server log and analyze them in great detail as a visit from a single visitor. This exercise will take you beyond the minutia of the detailed log files and show you how the pieces fit together to form a picture of what a reader's visit actually looked like. Next in Part 2 (Chapter 7, specifically), you're in for a treat—cookies. I'll explain what they are and how to bake them, and hopefully dispel some of the myths about them.

In Part 3, we will look at the commercial and freeware/shareware software that's out there for analyzing traffic. We'll also look at the service companies that analyze your traffic for you while you sit back and analyze their analyses. In this section we will also touch on system-administration functions you will want to consider for automating and generally making your traffic analyses as easy on yourself as possible.

To many folks, the only reason that traffic analysis is necessary is tracking and maximizing advertisement exposures. In Part 4, we will explore this strange new world. We'll take an overview of advertising on the web, then learn about how banner ads work, about ad servers, ad networks, and ad tracking and auditing services.

About the CD

The CD-ROM at the back of this book contains evaluation copies of some of the best of the field of commercial log analysis tools available today. In addition, the developers of these programs, and the two most important log analysis service companies, NetCount, LLC and I/PRO, have contributed HTML presentations to show and tell you why their particular package or service is best suited to your needs. Using the CD is easy. There's a file named Start.htm on the root directory. Just load this file into your web browser and follow the instructions and the links that you find in that document. If you need more explicit instructions to get started, see the page entitled "About the CD" at the back of the book.

About Program Listings in the Book

In several chapters of this book (especially Chapters 14 and 15) I have presented some simple CGI programs that I wrote in Perl. I'll be the first person to admit that I'm by no means an expert Perl programmer. But my intention here isn't to provide flawless program code. Rather, I want to convey how easy it is implement the functionality of these programs in any programming language. If you're a Perl programmer or if you intend to use the programs in this book, you may want to put a little polish on them first. For example, for simplicity's sake, I didn't take file locking into consideration. So the programs that open and write data to disk files will have problems if two users hit these files at the same time. Other than that, I think the programs are in pretty good shape. If you're a Perl purist, I'm sure you will call me on some inefficiencies like opening files twice, or perhaps not using the most effective algorithms. If this is you, chill out. Morgan (my tech editor) already beat me up for opening a file twice. But as they are, the programs are easy to understand—and they do work.

Keeping It Current

Yesterday I registered a new domain name with the Internic just for this book. The domain is websitestats.com. By the time you see this in print, I'll have a slick web site up and running just for you. I'll use it to keep you up to date on the latest and greatest log analysis packages and services. I'll post all of the Perl code for the programs in this book, and provide pointers to tools and resources on the web that relate to tracking web traffic and advertising. I'll provide software developers and service companies with a way to post rebuttals or dispute (or amplify) any comments I've made in this book. And, of course, I'll provide you with a way to contact me directly—or of course, you can always e-mail me directly at rick@rlsnet.com. But check out the web site. I'm sure you'll find it interesting. Have a cookie, and I'll see you in my server log files.

PART ONE

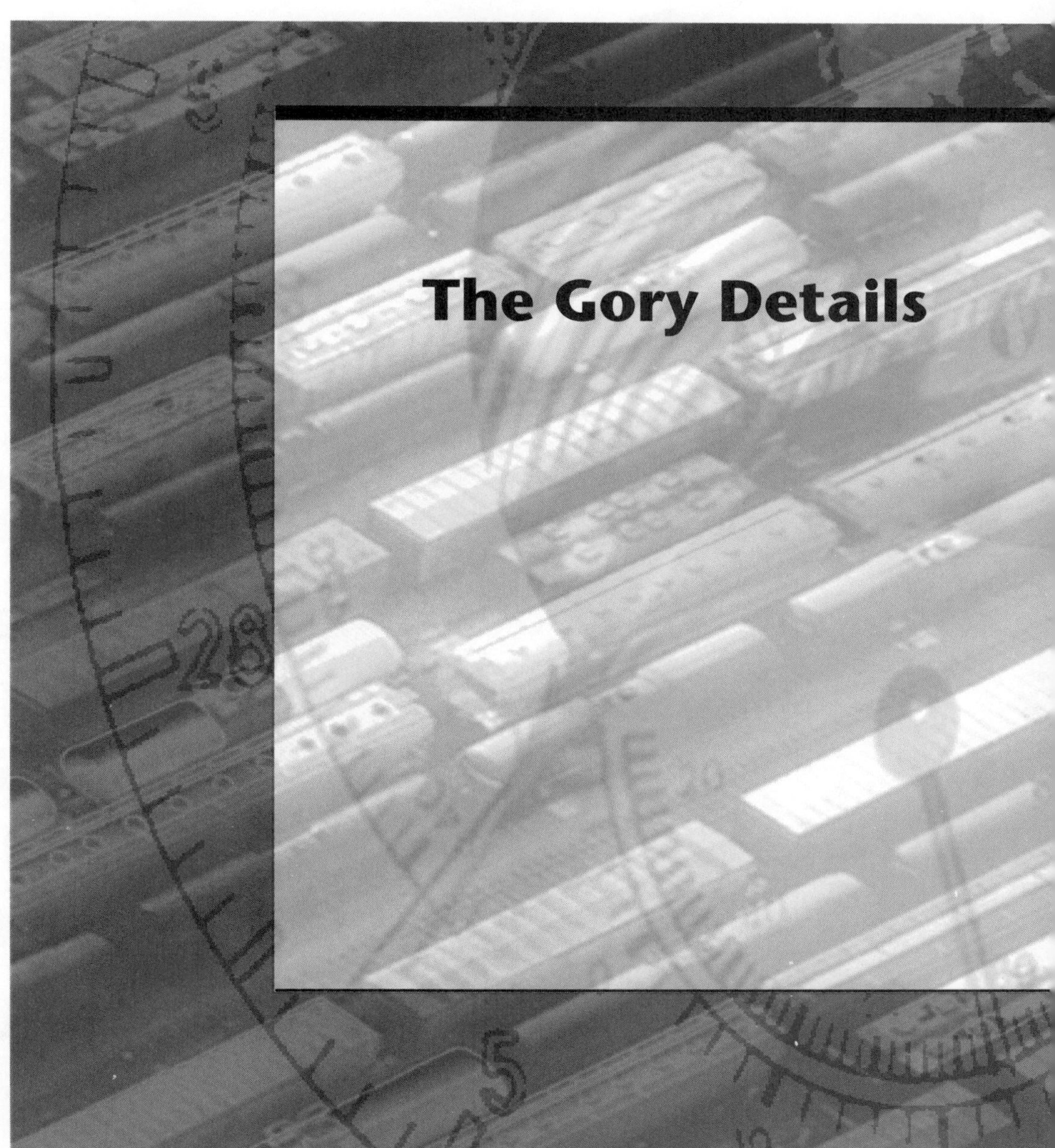

The Gory Details

Chapter One

The Basics of Tracking Web Traffic

One of the fastest growing niches for software developers is in programs that do high-level analysis on web server log files. The most comprehensive of these packages are integrated with sophisticated database systems that store detailed information about domains and networks on the Internet and other detailed data, to give you as much information as possible about who is using your web site and how they're using it.

Typically, these systems generate flashy reports complete with data tables, multiple "top 20" lists, and bar graphs and pie charts. Some packages prepare these reports for your color printer, while others actually create a collection of web pages to present the information through a web browser. The raw data that all of these systems use to do their magic is the same. They all start with the basic server log files. The developers of these packages do their best to shield users from the harsh realities of raw log file data; however, a fundamental understanding of those log files is key to knowing what it's possible to learn and deduce from the data that web servers record.

What better place to start a book about tracking web site traffic than at the very beginning? And the best place to begin is with the basics of the information that servers log as they do their work. In this first chapter, we will look at the building blocks for meaningful web site statistics—hits, views, and visits. Then, we will take a high-level look at each of the standard web server log files. In the remaining chapters of Part 1, we will examine each of the server log files in more detail.

If you're an experienced webmaster, you may have an uncontrollable urge to fast-forward to the good stuff: tracking and analyzing visits. If that's what you want to do, go right ahead—it is your book, after all (okay, it's *my* book, but you bought it), and it'll be a slow day in the daemon logs before I tell you how to read a book for which you laid down your hard-earned money. But even if you are a web server master, you may learn a thing or two, even in Chapter 1. If you're new to all this stuff, you'll definitely want to hang around for a solid grounding in web server log files. I promise to do my best to make this more interesting than reading the Internal Revenue Code.

Hits, Views, and Visits

If you've browsed the Web much at all, you've probably run across more than a few web pages sporting counters that claim something like: "1,764,345 hits since last week!" Should you believe them? Probably not. It shouldn't come as a surprise to you that the owners of many pages with counters like these have some sort of vested interest in the number of hits on their pages. Maybe they collect advertising dollars for each hit they draw, or maybe they just want to brag about the popularity of their web site. The trouble is, they never define for you exactly what they mean by a "hit."

When you pull up a web page in your browser, that single request can generate several transactions that get recorded in your log files. There's the original hit on the web page itself, and then there's a hit for each and every separate graphic image on that page; even a single request for a page with no graphics at all can generate more than a single hit. For example, let's say you type the following into a web browser's Location box:

```
http://www.interaccess.com/~charley
```

It's very likely that before the web server on the other end even begins to transfer Charley's home page to your browser, the server will generate a redirect to the more correct URL:

```
http://www.interaccess.com/~charley/
```

Notice the additional slash (/) at the end of the URL. Does Charley count that as a single hit or two? And after the page is loaded up and all 14 of the graphics on his page have been displayed, is the grand total 16 or 1?

Unfortunately, the Charleys of the Web seldom tell you how their counters are working, so you can't put much weight into what they say. To be fair to Charley, though, he may be using a counter that takes all of this into consideration and properly reports just one hit for such a view; then again, he may not.

On the other hand, it doesn't matter much at all what the Charleys of the Web put on their pages—at least as far as this book is concerned. What we're interested in here is the substance of the transaction: what really happened. You viewed Charley's home page. That's one *view.* We don't really care how many separate HTTP transactions had to occur for that one view to take place.

To continue our example, as you look over Charley's web page you might see a link to another of Charley's pages that interests you, so you click on the link. Another collection of hits is recorded in the web server's log files, and you view another page. This process repeats until you tire of Charley's pages and wander off to another web site. During your tour of Charley's web site, you may have viewed a handful of his pages. Your stroll through his pages is considered one *visit*.

Now put yourself in Charley's shoes. If you're trying to draw traffic to your web site, you probably want as much information as possible about who is visiting, where they came from (or how they found out about your pages), what path they followed through your web, and how long they lingered on each page that they viewed. The only way to know if anyone is visiting is to take a look at your server's log files.

At least, you may be interested in this level of detail at first, when your web site is brand new and drawing a crowd is of paramount concern. When masses of web users visit your site and you think you will drown in your log file data, you'll be looking for a different kind of relief: summary reports.

So, back to Charley for a minute: when his web pages are new and before he's drawn a great deal of traffic, every hit is likely to be interesting. As his pages become known and visited frequently, Charley's interest will probably transition quickly from the hit to the visit. The minute details of the paths that every reader follows just won't be as interesting as they were at first. Ultimately, as his pages endure many visits, Charley will get real tired of wading through megabytes of log file data to try to get an idea of how readers are using his web site. He'll seek relief in a software package or service company that will process the raw log files and create summary information that is concise and delivers what he wants to know.

Server Access Logs

In the next few chapters, we're going to look at each of the common web server log files in greater detail. But a short introduction to each here might be helpful.

There are actually two primary logs that every web server should use

- Transfer (or access) log
- Error log

While every activity is recorded in either the transfer or error log, that doesn't mean that all of the information available about a transaction is recorded in those logs. Two additional logs are common, although they're not necessarily widely used:

- Referrer log
- Agent (or user agent) log

Many web servers don't record these logs in their default configuration. And even if they do, many system administrators will turn them off. Once you're up to speed with tracking your traffic, you too will probably choose a log format that combines the information provided by the referrer and agent logs into the transfer log.

By the way, regardless of the type of computer you use (Unix, Windows NT, Macintosh, and so on), every web server for every platform records log files in basically the same way. The log files are plain text files—that is, they contain no special formatting characters or codes except for an embedded line feed (plus a carriage return in the case of an NT or Windows computer) at the end of each line. Each entry is a line of text appended to the end of the file.

The Transfer Log

The most important log file that your server records is the transfer log (which is sometimes called the access log). This is where you look to see the minutiae of what's going on with your server. Every hit on the server is recorded in the transfer log with the date and time of the request, the name or IP address of the computer from which the request came, the actual text of the request itself, a status code, and the number of bytes transferred to the requester. Actually, there's even more data in the transfer log, but we'll get into that in Chapter 2.

Here is an example of the transfer log entries that a single visit might generate:

```
cnx.conexis.com - - [28/Jun/1996:04:03:01 -0700] "GET / HTTP/1.0" 200 1586
cnx.conexis.com - - [28/Jun/1996:04:03:06 -0700] "GET /products/ HTTP/1.0" 200 2667
cnx.conexis.com - - [28/Jun/1996:04:03:09 -0700] "GET /art/bullet.gif HTTP/1.0" 200 920
cnx.conexis.com - - [28/Jun/1996:04:03:10 -0700] "GET /art/cactus.gif HTTP/1.0" 200 4288
cnx.conexis.com - - [28/Jun/1996:04:03:28 -0700] "GET /art/nav.jpg HTTP/1.0" 200 12781
cnx.conexis.com - - [28/Jun/1996:04:03:34 -0700] "GET /art/logo.gif HTTP/1.0" 200 15633
```

If you look at the text following the word "GET" in these lines, you can see the text of each request. In our example, the reader arrived at the top-level page (the root, or / page), then followed a link to the "products" page. In turn, the "products" page contains several graphics, each of which generated a hit in the transfer log.

The Error Log

When an HTTP transaction results in an error or failure status code, generally the server logs information about the transaction in both the transfer log and the error log. I say generally because not every result is recorded in both logs.

Among the errors that are logged in both files are requests for pages that don't exist, attempts to access files or directories for which a reader doesn't have permission, or any of a variety of possible error conditions.

Here are some lines from an error log:

```
[Wed Jun 19 16:26:52 1996] send lost connection to client gatekeeper.gwl.ca
[Wed Jun 19 12:22:19 1996] read timed out for garnier.thinice.com
[Thu Jun 20 06:57:00 1996] send lost connection to client sunlight.es.atl.sita.int
[Mon Jun 24 13:17:12 1996] send lost connection to client sfo-ca15-25.ix.netcom.com
[Wed Jun 26 14:01:52 1996] access to /www/docs/robots.txt failed for c.mv.opentext.com, reason:
        File does not exist
[Wed Jun 26 18:22:46 1996] send lost connection to client 165.133.59.20
[Thu Jun 27 16:58:33 1996] access to /var/www/docs/rlsnet/robots.txt failed for crawl2.atext.com, reason:
        File does not exist
[Fri Jun 28 06:32:47 1996] send timed out for ppp79.enter.net
```

Like the transfer log, the error log provides a time stamp for each line of the log file. This makes it possible to match up entries in the transfer log and error log, but it isn't easy. Notice above that the time stamp is in a different format in the error log than in the transfer log. We'll look more at these differences in Chapter 3.

One of the most common messages in the error log is the "lost connection" message. This is most often the result of a reader canceling a download in midprocess. Of course, this isn't really an error, but something that you may want to look at from time to time. In Chapter 2, we will look at what this type of error may mean, and what you can do about it.

The Referer Log

The referer log is one of the two optional log files. Virtually every server that offers one spells it the same way—wrong. Yes, it *should* be "referrer" (but who are we to complain?). The web server uses the referrer log to record the URL from which requests come. In other words, it records where your readers came from or how they found your pages. This is tremendously useful information when you're trying to build up traffic to your site, especially in the beginning. (As we mentioned earlier, once you've got as much traffic as you want, you may never again care where people are coming from.) But at least in the beginning, activate your referrer log or include the referrer information in your transfer log for some interesting reading.

The referrer log includes the URL from which the reader came and the page in your web where the reader arrived after following the link. Here's what one looks like:

```
http://search.yahoo.com/bin/search?p=nuts&a=n -> /index.html
http://search.yahoo.com/bin/search?p=apples&b=26 -> /index.html
http://search.yahoo.com/bin/search?p=arizona&b=76 -> /index.html
http://www.yahoo.com/Business_and_Economy/Companies/Food/Produce/Fruit/ -> /index.html
http://index.opentext.net/OTI_Robot.html -> /products/nuts.html
http://altavista.digital.com/cgi-bin/query?
       pg=q&what=web&fmt=.&q=Apple+Cider+in+Pennsylvanial -> /products/cider.html
http://www.altavista.digital.com/cgi-bin/query?
       pg=q&what=web&fmt=.&q=nutcracker ->/products/merch/nutcrack.html
http://www.azdirect.com/classads/food/food.htm -> /index.html
http://search.yahoo.com/bin/search?p=apple+cider&a=n -> /index.html
http://www.yahoo.com/Business_and_Economy/Companies/Food/Produce/Fruit/ -> /index.html
```

The full URL at the beginning of each line (together with search engine parameters in some cases) is the page from which readers came. The characters "->" represent an arrow—indicating that the reader came from the referrer page (on the left) and linked to the page on the right side of the arrow on the local web site.

When you use a separate referrer log to record referrer information, that's about all the information there is for you to know. There's no way to tie the entries back to specific visitors—or even to specific hits—unless you use a combined log format. We'll look more at combined log formats in Chapter 2, and at the referrer log in more detail in Chapter 4.

The Agent Log

Servers use the agent log (or user agent log) to record the name and version number of the user agent making the request. Most often, this is a web browser. But it's also common to find the names of robots and search engines lurking in your agent log. When your web site gets worked over by a friendly worm, it will leave its mark on your log files just as a casual reader will.

The most common entry in an agent log is some version of "Mozilla." Mozilla was the code name for the Netscape Navigator browser and remains a term of endearment for the product to this day—and the name that it reports to remote web servers. However, every browser that reports itself as Mozilla isn't necessarily a Netscape browser. In the following sample of an agent log, notice the line (about in the middle) as Mozilla/1.22 (compatible…). This is Microsoft's Internet Explorer browser (MSIE).

Here's a random sampling from the agent log of a production web server:

```
Mozilla/3.0b3 (Win95; I)
Mozilla/2.02Gold (Win95; U)
Mozilla/1.22 (Windows; U; 16bit)
MetaCrawler/1.2b libwww/4.0D
Mozilla/1.1 (Macintosh; U; PPC)
Mozilla/2.02Gold (Win95; U)
Mozilla/3.0b3 (Win95; I)
Mozilla/1.22ATT (Windows; U; 16bit)
Mozilla/2.0 (Win95; I)
Mozilla/3.0b4 (Win95; I)
Lynx/2-4-2  libwww/2.14
Infoseek Robot 1.17
Mozilla/3.0b3Gold (Win95; I)
Mozilla/1.22 (compatible; MSIE 2.0;  Windows 3.1) via Harvest Cache version 1.4pl3
Mozilla/2.01 (Win16; I)
Mozilla/2.0 (Win16; I)
OTI_Robot/2.0  libwww/2.17
aolbrowser/1.0 InterCon-Web-Library/1.1 (Macintosh; 68K)  via proxy gateway  CERN-HTTPD/3.0 libwww/2.17
CERN-LineMode/unspecified  libwww/unknown
Mozilla/2.0 (X11; I; BSD/386 uname failed)
Mozilla/2.02 (X11; I; SunOS 5.5 sun4c)
CERN-LineMode/unspecified  libwww/unknown
ArchitextSpider
Lynx/2-4-2  libwww/unknown
Mozilla/2.0GoldB1 (Win95; I)
MetaCrawler/1.2b libwww/4.0D
IWENG/1.2.000  via proxy gateway  CERN-HTTPD/3.0 libwww/2.17
```

Summary

Together, the four log files we have looked at briefly in this chapter constitute the basis of nearly everything we will cover in this book. We'll spend the rest of Part 1 taking a closer look at each of these log files—examining each field of each log to understand what they mean.

In the first chapter of Part 2 (Chapter 6), we will explore what more we can learn by bringing the data in each of these files together to present a complete picture of a visit. We'll also explore ways to link transfer log hits together to view them as visits—rather than an unassociated collection of log entries, and how to read your server statistics and take what you learn from them to make your web site more effective.

In Part 3, we will get into the goal of most of the first two parts—doing high-level analysis and generating statistics from server log files. We will take a detailed look at some of the best commercial log analysis software (most of which is included on the CD in the back of this book). We'll also look at log analysis service companies that will process your statistics for you automatically and virtually invisibly to you.

In Part 4, we will jump into a very important area for many people interested in web-based marketing—advertising. We will look at what people and companies are doing today to buy and sell web ad space, and at the mechanics of advertising, ad servers, ad networks, and ad tracking and auditing services.

Finally, we will take another look at tracing visits. We'll spend some time in Part 2 on tracking visits with cookies, but in Chapter 17, we will look at a completely different technology that promises to be interesting—if not downright exciting.

COLOR
FILM

Chapter Two

The Transfer Log

Every access to your web server is recorded in your server's transfer (or access) log. This includes both successful queries and errors generated for any reason at all. Each access, whether it results in a successful transfer or a failure, is considered a hit.

For every hit, your server adds a new line to the transfer log to record where the request came from, what was requested, the date and time, and other information. The most common format for a transfer log is called, sensibly enough, the Common Log Format.

In this chapter, we will explore in detail each and every field of the Common Log Format transfer log. We'll look at what each field means, explore the meanings of the values reported in those fields, and even second-guess the designers where they forged ahead without thinking about the ramifications of some of their decisions.

The Host Field

The first field of the Common Log Format transfer log is the host field. Most of the time, this will be the fully qualified name of the remote host making the connection and requesting a document; sometimes, it will be only the IP address of the client making the connection.

While you are still in the stage of looking at your log files regularly to see who's visiting your web, it's handy to be able to see the fully qualified domain name of the reader's computer rather than an IP address, which, without some additional legwork, has next to no meaning. For example, perusing your transfer log, which of these two log entries gives you more information?

```
pc36.cs.umn.edu - - [10/Jun/1996:18:28:25 -0700] "GET / HTTP/1.0" 200 5138
```

or

```
165.247.49.151 - - [10/Jun/1996:19:17:14 -0700] "GET / HTTP/1.0" 200 5138
```

Granted, taken individually, neither of these entries gives you very much information. But at least with the first one you can see that the request came from someone sitting at a PC (one named "pc36") in the computer science department of the University of Minnesota. This is a lot more information than you get with an IP address only. At least at a glance.

Eventually, however, you will probably want to turn off logging domain names—especially if your server draws any significant amount of traffic. The reason is that for your server to supply you with a name, it has to do quite a bit of work. It uses the Internet Domain Name System (DNS) to look up the name of the remote host. The data packets containing the request from the remote computer actually only contain the IP address of the remote user. So to give you a name instead, your system has got to make its own request to your network name server to resolve the IP address into a hostname.

Such a request normally takes only a few milliseconds on most networks, but on a heavily loaded server, these thin slices of CPU time and the additional network bandwidth can add up to perceptibly slower access times. Moreover, you don't lose any information in not doing a DNS lookup—you still have the IP address of the remote system. And the tools you will use to analyze your server statistics will likely be equipped to work with IP addresses as easily as they do with hostnames.

Just a couple more notes on the hostname field before we move on: A server configured to do a DNS lookup on every access and log hostnames will still fail to resolve IP addresses—often. This is because often a permanent hostname is not assigned to every IP address on the Internet that is allocated. One reason is that many Internet Service Providers (ISPs) assign IP addresses to their clients dynamically. This means that every time one of these clients fires up his computer to do a little web surfing, he will likely get a different IP address than the previous time he logged in. Service providers do this so that they can service a larger user base with fewer IP addresses.

What this means to you, however, is that you're not going to get a hostname in your transfer log—just an IP number. And when you get around to analyzing your log files, the best you can do is to get the name (and geographic location) of the ISP that owns that IP address.

Finally, you may be wondering how to configure your server to either do hostname lookups or not. Since I told you that you're probably best off if you disable hostname lookups, I would be remiss not to point you in the right direction. With the Netscape web servers, it's exceedingly easy. You just log onto your Administration server, choose Server Status and Log Preferences, then click either the Record Domain Names or Record IP Addresses radio button (see Figure 2-1).

With Unix servers, changing your option for logging domain names isn't quite as easy. With the NCSA server, you set a run-time option in your server configuration file. Typically, this file is named httpd.conf, and it's where you set up the most critical information that your server needs to run, such as the location of its document tree and log files. The keyword for the DNS mode is called DNSMode, and your choices are Maximum, Standard, Minimum, and None.

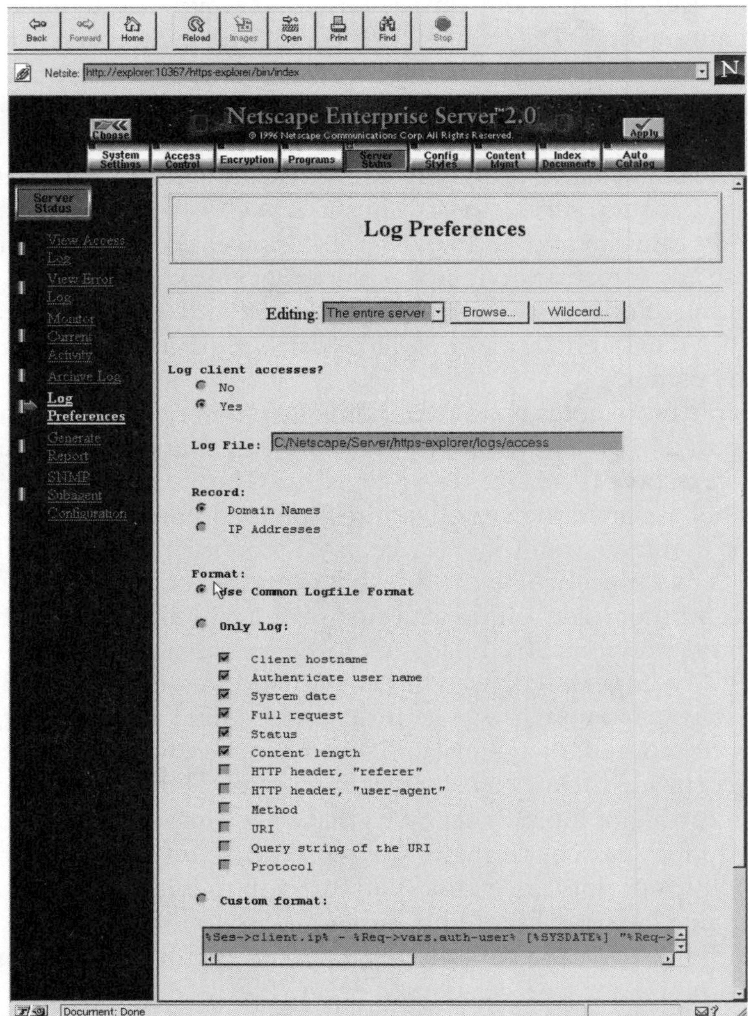

Figure 2-1. *Setting log options with Netscape's Enterprise Server*

The NCSA server (and the other Unix servers) are written to be portable to many different Unix platforms. In fact, there are so many different Unix platforms that it would be nearly impossible to offer binary versions ready to install on any machine. So system administrators download the complete source code to the servers and compile it on their own systems.

Unfortunately, some of the other Unix servers haven't offered the DNSMode run-time option yet. Until they do, the option of whether to log hostnames or IP addresses is something you have to decide and define when you compile the server. If you change your mind later on, you will have to change the option (probably in the makefile) and recompile the server.

The RFC 931 Field

The second field in the Common Log Format transfer log is variously known as the RFC 931 field, the ident field, or just the identification field. The most important thing to realize about this field is that it will almost always be simply a hyphen (-). That's because it's almost never used.

The intent of this field was to enable you, the webmaster, to know the actual identity of the person requesting one of your web pages (or at least the username). The trouble is that in order to get that information, the computer at the other end of the pipe must be running a program (yet another server daemon) that will respond to the request for the reader's name and supply it. There are so many obstacles in the way of this service that it's highly questionable whether it will ever make it into widespread use.

The original Request for Comment (RFC) on an authentication service was published in September of 1984. It was numbered RFC 912, and it laid out a proposal for a query/response protocol that would identify the owner of a TCP connection between two computers on the Internet. Here's an example of how it would work:

1. User wilbur on the machine tempest.northeastern.edu makes a TCP connection to another machine, sunny.usc.edu. Now, the web didn't exist in 1984, so the authors were thinking of connections to mail servers, FTP servers, telnet daemons, and so on. But you can think of this in terms of a request for a web page—Wilbur has requested a page at the University of Southern California.
2. As Wilbur's page is downloading, sunny.usc.edu contacts tempest.northeastern.edu on a completely different channel and says, "Hi tempest, I just got a connection from you on my port 80. Who made that request?"
3. Tempest will look at all of its open connections, find the one that sunny is asking about and responds with, "That request came from user wilbur." Then sunny thanks tempest, hangs up the connection, and logs Wilbur's name with the request.

In January 1985, the RFC was rewritten and renumbered (931), superseding RFC 912 completely. And in February of 1993, it was rewritten again and renumbered RFC 1413, making RFC 931 obsolete.

The title of RFC 1413 is *Identification Protocol*, and it defines the protocol for servers known as ident servers. Today, many Unix computers actually do run ident protocol servers—or at least they're equipped with them—whether or not system administrators choose to use them. The name of the actual program that implements the indent server is identd. The trailing "d" indicates that the program runs as a daemon.

The biggest obstacle to the ident field of web server transfer logs becoming useful is the relatively small percentage of computers connected to the Internet that actually run them. Unlike in 1984, and even 1993 for that matter, the majority of people using the Internet today do so from PCs running Windows 95 or a Macintosh. And they may not even have a direct connection to the Internet. Millions of web surfers access the Internet through information service providers like CompuServe and America Online.

> The Identification Protocol is not intended as an authorization or access control protocol. At best, it provides some additional auditing information with respect to TCP connections. At worst, it can provide misleading, incorrect, or maliciously incorrect information.
>
> —from RFC 1413

Windows 95 and the Mac operating system don't run ident servers, and so they don't respond to any ident requests. Technically, there's no reason they can't. But what compelling reason is there for anyone to go to the trouble of tailoring an ident server for Windows or the Mac? Giving others your name so their log files are a little more complete just isn't one.

You can also bet that the big information services like America Online are not going to provide any more information about their users than they have to. Security and privacy are both very real issues that are probably only deteriorated further with a protocol such as ident.

Nevertheless, requiring ident authentication is becoming more common in the Unix world. Some electronic mail servers are beginning to require ident authentication to accept e-mail messages from other systems. Even some FTP servers have been modified to authenticate FTP protocol connections to validate authorized users.

Whether or not the ident field will someday be useful for logging web server transfers is up for grabs. Until we know, expect the second transfer log field to be "-".

The Authuser Field

Most web servers implement a user authentication subsystem known simply as basic user authentication. Under this system, a web site administrator can choose to allow only explicitly authorized users access to certain directories in the web server's document tree. The administrator implements user authentication by placing a special file in subdirectories that are to be protected, creating a password database file that contains usernames and encrypted passwords and otherwise hooking it all together in the server's run-time configuration files.

Once a directory is protected this way, if a reader browsing your pages attempts to access any file in that directory, then the reader will have to successfully complete an authentication process before the server will oblige. Nearly all up-to-date web browsers know how to handle this procedure. The browser will pop up a dialog box with a message prompting the reader for a username and password (see Figure 2-2).

The reader will enter the username and password, click the OK button, and the browser will send the username/password data to the server. The server

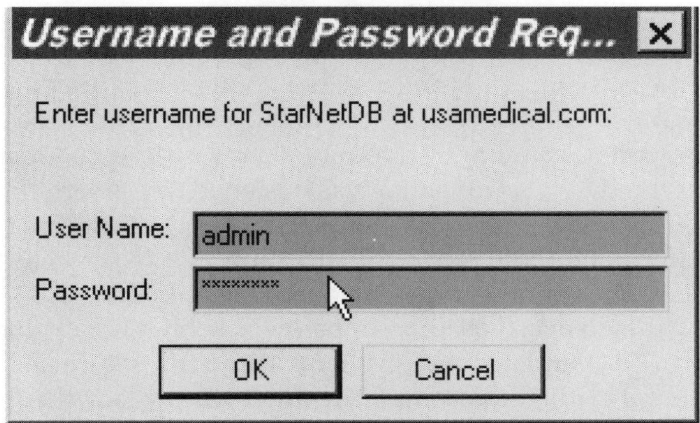

Figure 2-2. *The Netscape Navigator user authentication dialog box*

will check the username the reader entered against its password (or group) lists and encrypt the password sent by the reader using the same algorithm that was originally used to encrypt the password.

If the username is valid and the encrypted password matches, the server will serve up the page the reader originally requested along with an authentication key that the browser will retain and send back to the server in case the reader wants to access other protected documents. This way, the reader doesn't have to go through the authentication process for every document in the protected hierarchy.

This third field of the transfer log is called the authuser field. The authuser field gets the username (but not the password) that the reader enters. So, in a sense, this field is similar to the RFC 931 (or ident) field, except that the username isn't the reader's username on his or her own computer but the username for a protected portion of a web site—such as a database of confidential information. And, for this field, the reader has entered the username directly rather than the server determining it by going out the back door and around the side (as with the RFC 931 field).

Is Basic User Authentication Safe?

Yes and no. While basic user authentication will keep out the vast majority of unauthorized users, it's fairly quick work for a knowledgeable and determined hacker to figure out how to get in. Here's the rub:

When you type a username and password into a user authentication dialog box, these characters strings are not encrypted in a secure way by your web browser before it sends them to the web server at the other end. They are encrypted—but not by a routine that was designed for security. They're encrypted by a routine that was designed to convert binary data into an ASCII representation for transmission over communication mediums that can handle only ASCII characters. This routine is known as *uuencode*, and the corresponding decryption is done with *uudecode*.

To the untrained eye, a username/password combination encoded with uuencode will be completely indecipherable and indistinguishable from the rest of the gobbledygook around it. It won't look like anything readable to someone passively watching packets going by on a network. But if a hacker has targeted your site, it's not impossible to find uuencoded data, capture it, and uudecode it.

Presumably, any portion of a web site that an administrator would protect with user authentication might be an area of particular interest for watching who's actually using those areas. Having a concrete username to refer to can be very revealing—it even makes possible things like billing database users for their actual usage.

The Time Stamp

The least interesting, but nonetheless useful, field in the Common Log Format transfer log is the time stamp. There are actually three pieces of information in the time stamp:

- Date
- Time
- Offset from Greenwich Mean Time (GMT)

All three of these represent the local time for the server servicing the request, which brings up an interesting observation.

Since web servers aren't usually mobile—that is, they don't typically ride around in trucks or airplanes—why do they need to log the GMT offset for every access? Isn't this always going to be the same (except for a one-hour difference when changing between daylight savings time and standard time)? Isn't this redundant? How many terabytes of disk space are wasted worldwide because every web server's transfer log has to store these additional six bytes of data for each and every access? Who knows? (Perhaps the people who decided on the Common Log Format owned stock in a company that manufactured disk drives.)

All right, here's one conceivable use for the GMT offset. A particularly large web site may actually be a collection of mirrored sites that are geographically far apart. For example, a web site in Denver may have mirrors in London and Sydney to draw the brunt of the traffic from those continents. Assuming the administrators in Denver want to run statistics on the combined site, the GMT offset will become valuable for keeping the time zones straight.

Common Log Format specifies that the date is in the following format:

```
DD/Mon/YYYY
```

DD is the day of the month in numbers (preceded by a zero if it's the ninth day or earlier), Mon is the first three letters of the month name (e.g., Jun, Oct), and YYYY are the four numeric characters of the year.

A colon immediately follows the date, followed by the time in this format:

HH:MM:SS

HH is the hour, MM is the number of minutes past the hour, and SS is the number of seconds past the minute.

Finally, the last part of the time stamp is the GMT offset. A space character follows the time, then there is a plus or minus sign to indicate the number of hours the local time zone is from GMT (multiplied by one hundred for yet another obscure reason). For example, Pacific Daylight Time is seven hours west (or less than) GMT. Thus, the GMT offset as it is reported in the log file will be -0700.

Here's what an actual time stamp looks like:

```
[10/Jun/1996:18:28:25 -0700]
[10/Jun/1996:19:17:14 -0700]
```

These are from the two hits we looked at near the beginning of the section on the hostfield.

The whole time stamp could have been thought through better. It would have made the most sense to do away with the GMT offset completely and for the date and time to use this format:

YYYY/MM/DD:HH:MM:SS

Or, if there really is some value in the GMT offset, that could be tacked onto the end of this format. This would have allowed a pure ASCII sort to properly collate log data from multiple months. As it is now, alphabetically, Feb comes before Jan.

The HTTP Request

Now we arrive at the meat of the transaction—the actual request from the remote user. This is the fifth field in the Common Log Format transfer log. It's the only field that is surrounded with quotation marks, because it's the only field that can include embedded space characters (which everywhere else separate the fields).

The HTTP request starts off with a method keyword (or command, or request, depending on your point of view). The three possible methods are

- GET
- POST
- HEAD

The GET method tells the server that the reader is requesting a document or perhaps is specifying the name of a program that will do some processing and produce output that should be "given" to the client. This is the method used for a standard page request.

Most web browsers maintain a disk cache of recently viewed pages and graphics. If the browser determines that a requested page is in its disk cache, it will append a special header to the HTTP request, called an If-Modified-Since header. The data in this header is the time stamp of the file in the disk cache. This effectively turns a standard GET request into a conditional GET request. The server will check to see if the page or graphic has been modified since that time and will only send it if it has been modified. If it hasn't been modified, it sends a special status code back to the browser. You'll learn about that in the next section on the status code field.

The POST method tells the server that some data is following, and it's assumed that the URL given will be a program or script that can accept and make use of that data. When you fill in information on a form on a web page and click Submit, it's usually a POST action you're requesting.

The HEAD method is similar to the GET method, except that the server will only return the "<HEAD>" section of any document (along with the standard HTTP headers that are inherent in any HTTP transaction). This method is useful for testing the validity of hypertext links—usually in an automated way. (There's actually no way to make a standard web browser issue a HEAD request. With a browser, you either want the page or you don't. HEAD requests are most often generated by some other kind of program.)

The last part of the HTTP request field is the name and version number of the protocol. This is almost always HTTP/1.0, although occasionally you do see HTTP/0.9. Of course, now there's a new version of the HTTP protocol (version 1.1). You may begin to see references to the newer version soon. (Since web servers only respond to the HTTP protocol, one might wonder what the purpose is of sending "HTTP" back and forth all over the world.)

The Status Code Field

The penultimate field of the Common Log Format transfer log is the request status code field. This is a code that the server issues (both for writing to its log files and for responding back to the remote client). The status code describes the success or failure of the transaction.

Every status code is a three digit number, which is transmitted as three ASCII characters. There are four classes of status codes that are important to know about; they are listed in Table 2-1.

Status codes that fall into each of these four categories (or classes) have numbers in the same range. That is, all success codes are in the 200s, redirect codes are in the 300s, and so on. There is also a status class 100, but it's not used for anything yet. Thankfully, there aren't many status codes in HTTP version 1.0 (only 15 in all). However, version 1.1 of the HTTP protocol defines many more. See Appendix D for more details.

Table 2-2 shows the success codes. A code 200 is your basic successful retrieval of a web page, although this code is also returned from successful interactions with other entities, such as a CGI program. This is the most common status code you will find in your transfer log.

The 201 and 202 status codes are relatively rare. In fact, you may never see one because there's currently no widespread application for them. You will, however, see a code 204 on occasion. This usually indicates a problem with a browser or a CGI program improperly forming an HTTP header.

Table 2-3 presents the redirect codes. Code 301s aren't too common. A web site administrator can create an entry in the server's run-time configuration

Class	Purpose
200	Success
300	Redirect
400	Failure
500	Server errors

Table 2-1. *Status Code Classes*

Code	Description
200	OK
201	Created
202	Accepted
204	No content

Table 2-2. *Success Status Codes*

files (with the Redirect directive) that makes the server respond to a request for a specified URL with a code 301. This is supposed to notify the reader that the resource has moved permanently, but not all browsers support this. Those that don't support it display the familiar "This document has moved" message.

Code 302s, on the other hand, are very common. Your server will probably generate many of these on the fly. For example, in Chapter 1, I gave an example of a requested URL that lacked a trailing slash (/) on the end. Your server may automatically generate a redirect to the URL with the trailing slash added to the end.

Code 304s are also quite common. Whenever a user agent (a reader's web browser) has cached a document or graphic and the reader comes back to view that page again, the browser will submit a conditional GET indicating that the server doesn't have to send the document or graphic again if it hasn't changed. The code 304 is how the server responds to such a request, as if to say, "No, it hasn't changed, go ahead and use the copy you have."

Code	Description
301	Moved permanently
302	Moved temporarily
304	Not modified

Table 2-3. *Redirect Status Codes*

Table 2-4 shows the failure codes. The failure code that we all probably know the best is the 404 (Not found) status code. Every time you enter a URL wrong or follow a stale link, you'll see this familiar code. A 400 status code means that a reader has typed a URL so badly that the server couldn't even determine what the reader was asking for.

The 401 status indicates that user authentication is required to access the requested object, but that the user has failed to successfully authenticate himself or herself by entering a correct username/password pair (see "The Authuser Field" section earlier in this chapter).

A 403 status means that the server understands a request but it's not going to fulfill it. For example, if a user tries to obtain an index of the cgi-bin directory—or another directory for which the web server doesn't have adequate permissions—some servers will fail the request with a status 403.

Unfortunately, you can't count on web servers reacting in any uniform way to an error condition. For example with the Apache server (a popular Unix server), if you try to access a directory for which the server doesn't have adequate permissions, the server will issue a redirect (status 302) to nowhere instead of responding with a failure code. A redirect to nowhere results in your web browser timing out and issuing a DNS failure message. Go figure.

Table 2-5 shows the server error status codes. In a word, if you see one of these codes in your transfer log, you'd better put on your administrator's hat and find the problem—something's broken.

The thing to remember about errors (both failures and server errors) is that they won't be a material component in your web site statistics. They're more of an administrative red flag to show you that something could be wrong with your server configuration or CGI programs or that someone's trying to access objects they shouldn't be.

Code	Description
400	Bad request
401	Unauthorized
403	Forbidden
404	Not found

Table 2-4. *Failure Status Codes*

Code	Description
500	Internal server error
501	Not implemented
502	Bad gateway
503	Service unavailable

Table 2-5. *Server Error Status Codes*

If you're really interested in tracking the minutiae of every error, the best policy is to know your particular web server and how it reacts to each possible type of error. Then, when you analyze your log files (or prepare them for automated analysis), you'll have a better understanding of the information that's there.

The Transfer Volume

The final field of the Common Log Format transfer log is the transfer volume. This is an ASCII representation of the number of bytes transferred by the server to the client as a result of the HTTP request.

Since the only transactions that transfer any data at all are GET requests that result in a status 200 (OK), the transfer volume field isn't applicable on any other log lines and will contain only a hyphen (-) or a zero (0). For example, if a user agent makes a conditional GET request for a document that has not been changed, the server will log a status 304 (Not modified) and the transfer volume field will be simply a hyphen (-). A successful HEAD request, on the other hand, will result in a status 200 (OK) but will display zero (0) bytes transferred.

Optional Transfer Log Fields

In Chapter 1, I mentioned two additional common log files that supplement the information provided by the transfer log. These are the referrer and agent logs. The referrer log records the URL of a web page or resource that a user followed to arrive at your web site. This could be a page on another web site with a hypertext link to a page on your site, or it can even be the URL of a search engine together with the search criteria that a reader entered to find a listing for your web site. The agent log records the web browser name and version number (and sometimes the operating system) that a reader uses. (You will read more about these two logs in Chapters 4 and 5.)

A popular deviation from the Common Log Format is called the combined, or extended, log format. The combined log format integrates the referrer and agent data into the transfer log. Of course, once you do this, your transfer log is no longer in Common Log Format, but the combined log format may be more efficient and desirable to use. For one thing, with the combined log format, you can match up referrer and user agent information to the hits to which they relate.

Any deviation from the Common Log Format will break some log analysis software. But increasingly, log analysis software developers are making their products forgiving of extended log formats—either by supporting them directly, or by making them ignore fields that they can't interpret. For example, with an extended log format (any extended log format), the first seven fields of the log file are the same as the Common Log Format. Any additional fields are just tacked onto the end of each line. For this reason, it should be easy for software developers to get the data from the first seven fields and ignore the rest (if they don't intend to support the extended fields directly).

As I write this, several of the commercial software analysis packages still insist on pure Common Log Format. We'll go into which ones in detail in Chapter 9, but if by the time you read this, they don't support extended log formats directly, you might want to move on to the next package on your list. In Chapter 7, we will explain in detail about something called server-set cookies. If you're interested in tracking your site traffic as accurately as possible, you're going to want to implement cookies. This will require using an extended log format, more than likely one that includes three additional fields beyond the Common Log Format. These are the referrer, user agent, and cookie fields. But you won't get much mileage out of using cookies if your analysis software doesn't support them being in your transfer log.

In Chapters 3, 4, and 5, we will look at the other three common log files: the error log, the referrer log, and the agent log. Then, in Part 2, we will look at how to bring together all of the data from each of these log files to form a coherent picture of the activity at your web site.

Chapter Three

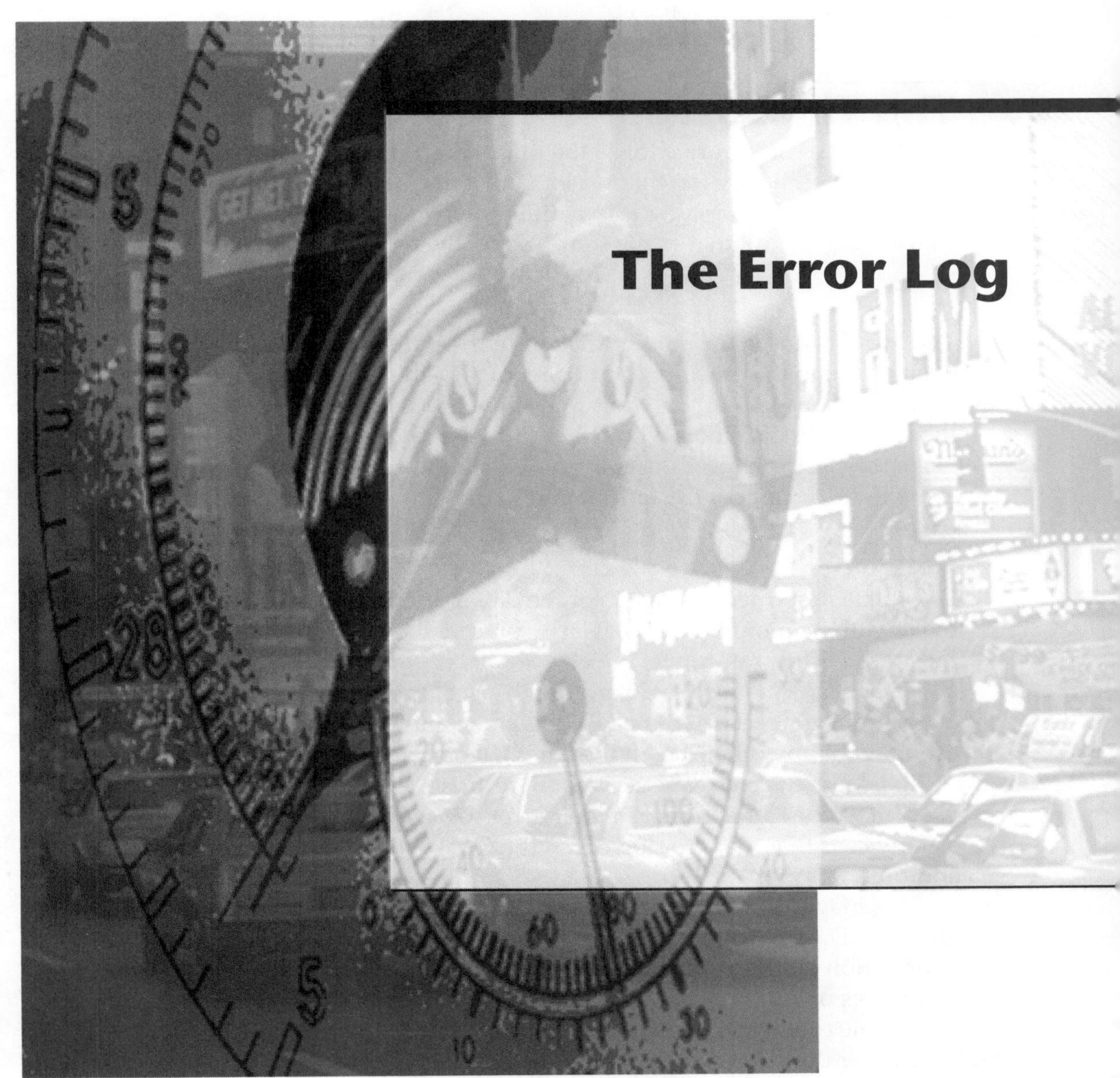

The Error Log

The error log gets a new log line for each failure or error condition that the server encounters. Essentially, the error log consists of two fields: a time stamp and a description of the error or failure. In this chapter, we will look at both of these fields. First, we'll complain just a little more about time stamp formats, then we will take a look at the types and formats of errors that are documented in the error log.

The Time Stamp

The first field of the error log is the time stamp. This is similar to the time stamp in the transfer log that we looked at in Chapter 2, but, inexplicably, the time stamp in the error log is in a different format than the time stamp in the transfer log. Recall that in the transfer log, the time stamp is in this format:

```
[DD/Mon/YYYY:HH:MM:SS -0000]
```

The format chosen for the time stamp in the transfer log was pretty bad. It's even worse in the error log. But worst of all—they're not even the same!

For reasons known only to some aging graduate student/programmer at the University of Illinois who either didn't take the time to check on the format his partner used in the transfer log or was endowed with a streak of particularly dry wit, the format of the time stamp in the error log is the following:

```
[Day Mon D HH:MM:SS YYYY]
```

Day is the first three characters of the day of the week (e.g., Sat, Wed) and D is the day of the month (not necessarily preceded by a leading zero if it's before the tenth of the month).

This means that if you do an ASCII sort of the contents of your error log, every entry that occurred on a Friday will come before any entry created on a Monday (or any other day of the week), regardless of the week or the month.

To be fair, the Netscape servers (Enterprise Server and Fastrack Server) use almost the same format in their error log as in their transfer log. The only difference is that the error log drops the GMT offset and reports only the date and time.

This may bring up another question: how can different servers use different formats? The Common Log Format that we looked at in depth in Chapter 2 defines only the contents and format of the transfer log. There are no strict guidelines or rules for the format of the other log files. Moreover, even if there were some rules of the road, there's no one to enforce them. The only guidelines that server developers feel compelled to follow are the expectations of their users. Since there were so many tools early on to analyze transfer logs—tools that depend on the transfer log being in a consistent format—developers felt obligated to stick with the Common Log Format. But many (if not most) log analysis tools don't operate on the error log—so it seems that server developers haven't felt the need to get together on a standard.

Error Log Messages

An error line is generated in the error log for any of the error or failure result codes that we looked at in Chapter 2, plus a few additional ones. In the error log, errors are reported not by code (as they are in the transfer log) but rather by a textual description of the error. Thus, one way to think of the error log is as a more readable version of the errors reported in the transfer log. Unfortunately, the different time stamp formats make it difficult to match up the entries.

Error messages in the error log fall into several categories: administrative messages, access failures, lost connections, and time outs. The first category we will look at are administrative messages. Since these have no bearing at all on your site statistics, we'll look at them first to get them out of the way.

Administrative Messages

Administration messages are informational text that your server logs, which have nothing to do with readers accessing your site. These include

- Server startup messages
- Signals the server receives
- Server shutdown messages

When you start up your server for the very first time, one of the first things the server will do is create the server log files. Initially, they will all be empty—but not for long. Before the server settles in to wait for HTTP connections, it will use the error log to report that it started up successfully (or not). The Apache server's startup line will look something like this:

```
[Sat Jul  6 13:06:45 1996] Server configured -- resuming normal operations
```

The NCSA server's startup message is similar:

```
[Sat Jul  6 15:43:02 1996] HTTPd: Starting as ../httpd -d /var/www/
```

When the server receives a signal from the operating system, such as the signal to stop the server or restart itself, it will log both the receipt of the signal and any action it takes as a result of the signal. For example, here's the Apache server:

```
[Sat Jul  6 13:07:32 1996] SIGHUP received.  Attempting to restart
[Sat Jul  6 13:07:32 1996] Server configured -- resuming normal operations
```

Here's the NCSA server:

```
[Sat Jul  6 16:46:38 1996] HTTPd: caught SIGHUP, restarting
[Sat Jul  6 16:46:38 1996] HTTPd: successful restart
```

Unfortunately, if something is amiss in your server configuration files and the server can't start up normally, you can't necessarily count on receiving an adequate description of the problem in the error log. This doesn't have anything to do with your site statistics—but here's a tip: try starting the server from the command line. This way, you can see the standard output and standard error from the server as it attempts to start up. More than likely, you'll get a more descriptive error message than those logged to the error log.

Access Failures

The remaining error types fall into two categories: access failures and lost connections. Access errors correspond directly to the status code class 400 that we looked at in detail in Chapter 2. Since the 404 (Not found) status is the most common that you will run across, let's look at it.

Recall that if a reader makes a request for a page that doesn't exist, the server will return a status 404 to that reader's web browser as well as log the failed attempt in the transfer log. The transfer log entry looks this:

```
winpc.crs.com - - [08/Jul/1996:09:55:00 -0700] "GET /events.html HTTP/1.0" 404 -
```

The status code 404 indicates that the requested file (/events.html) does not exist, and the hyphen in the final field indicates that no bytes were transferred.

In addition to adding the above line to the transfer log, the server adds a corresponding line to the error log. This entry looks quite a bit different:

```
[Mon Jul  8 09:55:00 1996] access to /var/www/docs/events.html failed for
winpc.crs.com, reason: File does not exist
```

Note that in the error log this is actually one long line. Obviously, we have to wrap it here so you can see the whole line.

The first thing you may notice is that the time stamp is the first field of the log, as opposed to the hostname, which is the first field in the transfer log. The rest of the field is essentially one long sentence describing the nature of the error. It's not a particularly terse sentence—it's even downright verbose. Again, this is because the error log was originally designed primarily for human eyes as opposed to a computer format that would be optimum for a computer program to read. In this first example—the error log entry for a status 404 (Not found) error—the descriptive sentence is the following:

```
access to </path/file> failed for <host>, reason: File does not exist
```

There are essentially six common reasons for an access failure. Table 3-1 presents the most common error reasons you will find in an error log along with their corresponding status codes in the transfer log. For each of these failures, the first part of the error message is always the same, so you can say generally that access failures are logged in this format:

```
access to </path/file> failed for <host>, reason: <reason>
```

Lost Connections

Another common incident that shows up in the error log is lost connections. Most often, these are the result of readers canceling their download of a page by clicking either the Stop button or the back arrow on their browser, by clicking on a link in the new page before the page and all of its graphics are

Transfer Log Status Code	Error Log Reason Description
401 (Unauthorized)	user <user> not found
402 (Unauthorized)	user <user>: password mismatch
403 (Forbidden)	attempt to invoke directory as a script
404 (Not found)	file does not exist
405 (Not found)	script not found or unable to start
406 (Internal server error)	malformed header from script

Table 3-1. *Common Error Log Reasons*

completely transferred, or even by closing their web browser in the middle of a transfer.

Most web servers don't log a lost connection event in the transfer log at all. This error shows up only in the error log. We'll look more at the significance of this later in the chapter.

Although servers don't create a separate entry in the transfer log for a canceled transfer, the transfer log may still give you a clue to some cancellations. A GET request that is canceled in midtransfer will be logged a status code 200 (a normal successful transfer), but the number of bytes transferred (in the last field of the transfer log) may be less than the actual number of bytes in the object. However, some cancellations happen too late in the process for the server to do this—after the server has completed its transfer, logged it, and the host computer is just waiting for the outbound buffers to be emptied.

Connections can be lost either during a request before the server has even begun to process the request or after the request has been completed and the server is in the process of sending the requested documents. A connection lost during the request stage is logged as a "request lost connection." A connection lost during the transfer stage is logged as a "send lost connection."

The general format for a lost connection entry is the following:

```
<send|request> lost connection to client <host>
```

Here are a few entries from an actual error log:

```
[Thu Jun 13 13:52:00 1996] request lost connection to client 206.42.98.113
[Thu Jun 13 13:58:56 1996] send lost connection to client 206.42.98.113
[Fri Jun 21 09:13:23 1996] send lost connection to client interlock2.mgh.com
```

Time Outs

The situations that cause time out messages can vary widely from server to server. They're nowhere near as common as a lost connection, but they do show up occasionally. Usually, a time out error is the result of the server having to wait on some device or service other than the client that made the connection.

For example, a problem with a disk drive or an NFS-mounted file system may manifest in a time out error. Another example is the physical connection to the Internet being broken as when a reader's modem drops its line.

Useful Information in the Error Log

When you think about how server log files have probably evolved over time, the human readability of the error log makes some sense. If the transfer log were the only server log, administrators would have to wade through megabytes of log data to find errors, and then they would have to look up the error codes and otherwise decrypt the terse information. It makes sense to have another log that captures just the errors—and to record them in a way that makes it easy for administrators to stay on top of any problems on their web site. If that were the extent of the differences between the transfer log and the error log, you could essentially ignore the error log as being a useless duplication of the information in the transfer log (at least as far as site statistics are concerned).

However, the information in the error log is not merely a verbose duplication of information already in the transfer log. There is some additional information in the error log that is not logged at all in the transfer log—information that can be very useful for tailoring a more effective web site. This is in the log entries for lost connections—which are usually triggered by a reader canceling a download in midprocess.

If you've read many books about the World Wide Web, you will likely want to scream if you have to read one more author cautioning you to "be considerate of low-bandwidth users" by keeping your graphic files small. It's also likely that nothing you've ever read on the topic has explained how to actually measure the frequency of readers who get bored waiting for your graphics to download and bail out.

The server error log supplies this valuable nugget of information. Unfortunately, the format of the error log (being designed for your eyes rather than your analysis tools) makes it harder than normal to extract this information.

In Chapter 6, we will extract some of this data from the error log and integrate it with the data from the other log files to help create a more complete picture of how your web site is being used, and which pages may in fact be too graphic-intensive—causing your readers to abort transfers before they are complete.

Increasingly, server developers are modifying their servers to log connections in the transfer log indirectly. They are implementing this by logging only the actual count of bytes that are transferred rather than the total number of bytes contained by the requested object. As I write this, at least one of the log analysis software developers and one of the major analysis service companies are taking this into account. As this logging technique becomes more common and standard, we will probably be seeing all of the log analysis packages extracting this information from the transfer log.

Chapter Four

The Referrer Log

In the last two chapters, we looked at the primary log files that web servers use to log requests and errors—the transfer (or access) log, and the error log. In this chapter, we will examine the referrer log—the first of the two ancillary log files. Both the referrer log and the agent log (which we will look at in Chapter 5) are very simple log files. There's not a great deal to learn about them except that you shouldn't use them at all if you can get away with it. Read on to find out why.

The Referrer Log Format

The referrer log consists of only two fields of information separated by a symbol that looks like an arrow (a hyphen and a greater-than symbol surrounded by two space characters). The fields on either side of the arrow are the following:

- The URL of the page a reader's browser was displaying immediately before it accessed the path and filename (see next item).
- The path and filename of the page in your web site's document tree that the reader viewed.

This means that when readers are clicking around inside your web site, the left side of the arrow is the URL of the page on your site that they clicked on, and the right side of the arrow is the path/filename of the page where they went when they clicked on the link. This isn't terribly useful information when the origin and destination are both within your own site, because you already have this information (and in much more detail) in the transfer log. For example, look at this referrer log entry:

```
http://cidermill.com/info/ -> /info/aboutus
```

This is another representation of these lines in the transfer log:

```
usr.az.us - - [12/Jun/1996:10:19:46 -0700] "GET /info/ HTTP/1.0" 200 1776
usr.az.us - - [12/Jun/1996:10:20:26 -0700] "GET /info/aboutus HTTP/1.0" 200 2915
```

In the referrer log entry, you can see that a reader followed a link on the "info" page to the default page in a subdirectory named "aboutus." That's the extent of what you can learn from the referrer log entry. The two corresponding lines recorded in the transfer log give you the same information but in much more detail. Actually, the two transfer log lines don't explicitly tell you that the reader followed a link on the "info" page to the "aboutus" page, but you can safely infer this. After all, the "aboutus" page was the next page requested after the "info" page—and how else would the reader have known about the "aboutus" page?

On the other hand, when a reader enters your site for the first time, the referrer log effectively shows you where the reader came from. Now this is very useful information. For example, consider the following referrer log entry:

```
http://search.yahoo.com/bin/search?p=Apples&a=n -> /index.html
```

This entry shows you that the reader submitted some search criteria to the Yahoo search engine. The result of the search was a page sent to the browser by the Yahoo engine like the one in Figure 4-1.

When your reader clicks on the link to your site, the reader's browser sends the last URL to your web server so that it can log that information in the referrer log along with the page that was the result of the search. In this case, the search brought the reader to the /index.html page, the top-level page of the site.

Confidentiality and Referrer Logging

When a reader follows a link on a web page to another web page, the referrer information (the URL of the page containing the link) is encoded in the HTTP request header fields. (If you're interested in the details of the HTTP header fields, check out Chapter 7 and Appendix C.) However, sending the referrer information is a function of the reader's web browser, and whether or not a web browser sends referrer information is optional—at least in terms of the HTTP specification. If a web server receives an HTTP request without any referrer information, it will happily service the request—it just won't log any referrer information, because there isn't any.

The HTTP protocol specification strongly suggests that user agents (web browsers) make sending referrer information optional for the end user. After all, they may wish to remain anonymous, and the pages from which they arrive at a web site may be private. Unfortunately, browser developers have largely ignored this and provide no configuration option for a user to turn off sending referrer information.

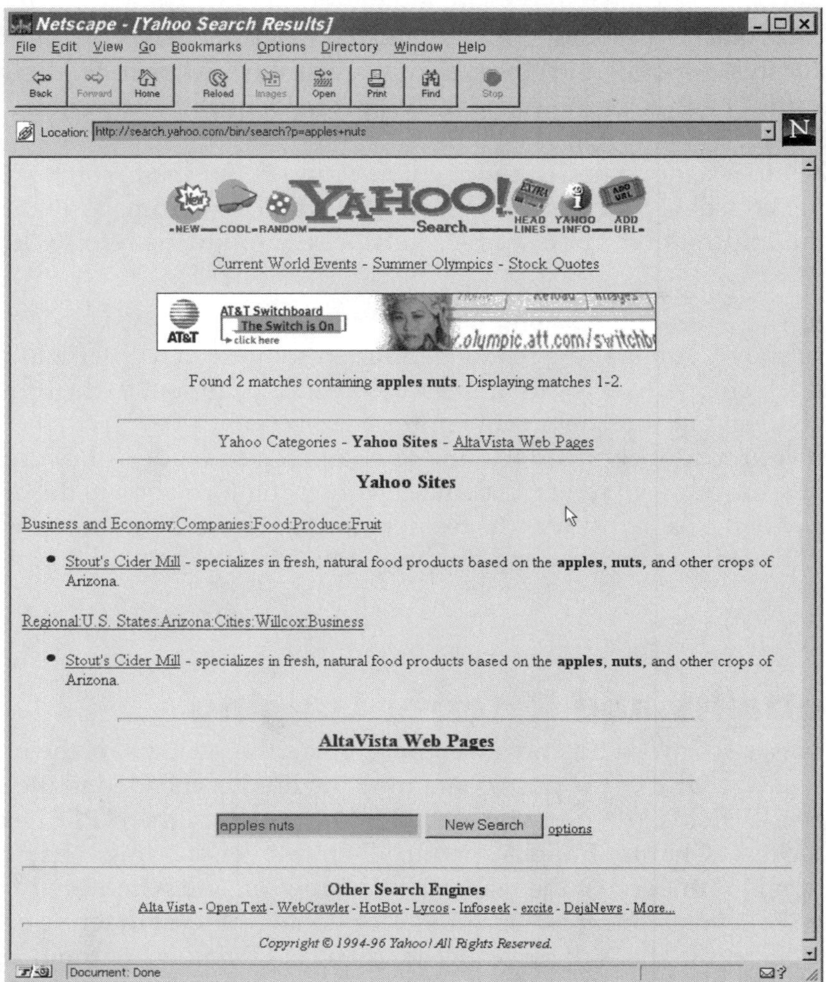

Figure 4-1. *The result of a search for a word on the Yahoo search engine is a page that displays links to sites (or individual web pages)*

On the other hand, if your web site statistics are your primary interest, you will probably appreciate that users can't choose to disable sending referrer information. If people could browse the web anonymously, most people probably would. And you wouldn't have nearly as much information about where your readers come from.

Tailoring Referrer Data

If you choose to log referrer data, then every time a reader follows a link on your web site to another page on your web site, an entry will be generated in the referrer log. As I mentioned earlier, because these entries are essentially duplications of data you will have in the transfer log (and not very useful when both the origin and the destination are within your own site), you may choose not to log referrer data within your own site.

Web server developers give you a way to stop logging referrer data for domains or parts of domains. With the Unix servers, you use the RefererIgnore directive to make the server stop logging referrer data for a whole domain or part of one. Check the documentation for your particular server for the proper syntax. Basically, to stop logging referrer information for your own domain, you just include a line similar to the following in your httpd.conf file:

```
RefererIgnore cidermill.com
```

(Of course, you would replace cidermill.com with the name of your own domain.) This will cause your server to log referrer information only for HTTP requests that don't have "cidermill.com" in the request header.

Avoiding the Referrer Log

Many web servers also allow you to combine referrer data with the transfer log. In Chapter 2, we mentioned the combined log format, in which the transfer log has two additional fields: a referrer field and a user agent field.

Combining the referrer information with the transfer log actually makes a lot of sense. If you log referrer information separately in the referrer log, you don't have any time stamps or hostnames to match entries up with their corresponding entries in the transfer log. If you log the referrer data within the transfer log, however, then the referrer data is logged right next to the relevant transfer log data.

The data recorded in both the referrer and agent logs are best reported in the combined log format. The trouble is that once you switch to the combined log format, your transfer log is no longer in the Common Log Format. And a transfer log in a format other than the Common Log Format will break some log analysis software (but not the better ones).

If you plan on developing your own log analysis tools, you should be thinking about combining all of the data you can into the transfer log. Sure, the line lengths will be long, but the human readability of the transfer log should be at the bottom of your list of priorities.

If you plan on using a shareware or commercial software package to analyze your log files, find out up front whether or not it will work with the combined log format. Some packages will simply ignore the extra fields (and not make use of the referrer or agent data). Others will require, or strongly suggest, that you use a combined format—these are probably the most desirable analysis tools. Still others won't work at all if the transfer log isn't in strict Common Log Format.

In the next chapter, we'll look at the final ancillary log, the agent log. Then we'll move into Part 2 and start to get into some of the good stuff—like analyzing your server log files and tracking visits.

Chapter Five

The Agent Log

In Chapter 4, we looked at the first of the two ancillary log files, the referrer log. In this chapter, we will examine the second one: the agent log. Like the referrer log, the agent log is simple; in fact, the agent log is even simpler than the referrer log. And like the referrer log, the user agent information is probably best logged in the transfer log in the combined log format—that is, if you want to be able to tie user agent data back to specific hits and visits in the transfer log.

The Agent Log Format

Technically speaking, the entire content of a user agent log entry is a single string of characters sent by the remote web browser in a request header field of its HTTP requests. This string of characters is defined by the browser developer and is hard-coded into the browser program. But even though agent log strings are defined by browser developers, you can count on a few standards here.

The first characters appearing in an agent log entry form a word—the name of the web browser the reader is using. By far the most common browser name you will find in an agent log is Mozilla—Netscape's pet name for its Navigator browser. Other common browsers are Microsoft Internet Explorer, IBM WebExplorer, and Lynx.

Don't assume that every line in an agent log that contains the word Mozilla was created by Netscape. Some browsers (Microsoft's Internet Explorer, for one) record themselves as Mozilla to show the Netscape version with which they are compatible. For example, Internet Explorer will record this: Mozilla/2.0 (compatible; MSIE 3.0; Windows 95).

Besides browser names, you'll also find the names of spiders and worms. No, not black widows and night crawlers, but programs that traverse the Net, following links and exploring domains to catalog pages they find and list them in their databases. A few of these programs (which are also called robots) that will work over your site are ArchitextSpider, InfoSeek Sidewinder, and MetaCrawler.

Immediately following the browser name is a slash character (/) and the version number of the software. Next there's a space character and another string of text, which you can essentially consider the second field of an agent log entry. This second field is surrounded by parentheses and contains the name of the operating system that the reader's browser is running on and sometimes some additional information about the computer. In a nutshell, you can say that the format of an agent log entry is generally the following:

```
<browser name>/<version> (<operating system>)
```

In addition to this information, you will sometimes find proxy servers inserting or appending their own name and version number into or to the end of an agent log line. This happens when a reader is reading your pages from a computer on a network that uses a proxy server. The proxy server interprets the reader's browser's HTTP request, makes the request itself on behalf of the reader's browser, and tacks its own plug onto the end of the HTTP request header. This is discouraged by the HTTP Working Group at the W3C Consortium (the people who write the HTTP specification), who say that doing this "makes machine interpretation of the field ambiguous." But proxy developers seem to ignore it—probably because of the ambiguity of that statement.

Useful Information from the Agent Log

A major problem with the agent log is that there is no way to tie the entries in it back to individual hits in the transfer log. There is no hostname, not even a time stamp, so the data recorded in an agent log has to stand on its own. You can still do some useful analysis with an agent log, but you just can't match it up with data in the transfer or error logs (not easily, anyway).

Let's look at an actual agent log. Here are a few lines of entries from a working agent log:

```
Lynx/2.5FM  libwww-FM/2.14
Slurp/1.0 (http://www.inktomi.com/slurp.html)
Lynx/2.3 BETA  libwww/2.14
     via proxy gateway  CERN-HTTPD/3.0 libwww/2.17
Mozilla/2.0 (Macintosh; I; 68K)
Mozilla/2.0 (Macintosh; I; 68K)
Mozilla/2.0 (Macintosh; I; 68K)
Mozilla/2.0 (Macintosh; I; 68K)
Mozilla/2.0 (Win16; I)
Mozilla/2.0 (Win16; I)
Mozilla/2.0 (Win16; I)
Mozilla/2.0 (Win16; I)
Mozilla/2.0 (Win16; I)
MetaCrawler/1.2b libwww/4.0D
MetaCrawler/1.2b libwww/4.0D
MetaCrawler/1.2b libwww/4.0D
Mozilla/1.22 (compatible; MSIE 2.0; Windows 95)
Mozilla/1.22 (compatible; MSIE 2.0; Windows 95)
Mozilla/1.22 (compatible; MSIE 2.0; Windows 95)
```

```
Mozilla/1.22 (compatible; MSIE 2.0; Windows 95)
Mozilla/2.02 (Win95; I)
Mozilla/2.02 (Win95; I)
Mozilla/2.02 (Win95; I)
Mozilla/2.02 (Win95; I)
Mozilla/2.0 (Macintosh; I; PPC)
Mozilla/2.0 (Macintosh; I; PPC)
Mozilla/2.0 (Macintosh; I; PPC)
Mozilla/2.0 (Macintosh; I; PPC)
Mozilla/2.0 (Macintosh; I; PPC)
aolbrowser/1.1 InterCon-Web-Library/1.2 (Macintosh; 68K)
     via proxy gateway  CERN-HTTPD/3.0 libwww/2.17
aolbrowser/1.1 InterCon-Web-Library/1.2 (Macintosh; 68K)
     via proxy gateway  CERN-HTTPD/3.0 libwww/2.17
aolbrowser/1.1 InterCon-Web-Library/1.2 (Macintosh; 68K)
     via proxy gateway  CERN-HTTPD/3.0 libwww/2.17
aolbrowser/1.1 InterCon-Web-Library/1.2 (Macintosh; 68K)
     via proxy gateway  CERN-HTTPD/3.0 libwww/2.17
```

As expected, you see a lot of entries for Mozilla. There are also entries for Lynx and America Online's browser (aolbrowser). An analysis of this log file with a little over 4,000 hits showed that 94 percent of all hits were by visitors using some version of Netscape Navigator; among those using Navigator, 84 different versions of the software were used. That may seem like an unbelievable number of versions of Netscape, but when you consider the many different operating systems that run Navigator (NT, Windows 95, Macintosh, Sun, other Unix computers), Navigator versions ranging from .9 to 3.0, and subcategories of operating systems like 16-bit and 32-bit versions of Windows, then the total can grow quickly.

The remaining 6 percent of hits are split among other web browsers and web robots. They are (in descending order of their frequency with the number of occurrences of each):

```
47 Slurp/1.0 (http://www.inktomi.com/slurp.html)
40 CERN-LineMode/unspecified  libwww/unknown
22 aolbrowser/1.1 InterCon-Web-Library/1.2 (Macintosh; 68K)
          via proxy gateway  CERN-HTTPD/3.0 libwww/2.17
22 MetaCrawler/1.2b libwww/4.0D
15 NetCruiser/V2.1.1
14 IBM WebExplorer DLL /v1.03
11 /0.5 libwww-perl/0.40
 9 IWENG/1.2.003  via proxy gateway  CERN-HTTPD/3.0 libwww/2.17
 8 Lynx/2-4-2  libwww/unknown
 8 GNNworks/v1.2.0  via proxy gateway  CERN-HTTPD/3.0 libwww/2.17
 6 aolbrowser/1.0 InterCon-Web-Library/1.1 (Macintosh; 68K)
          via proxy gateway  CERN-HTTPD/3.0 libwww/2.17
 5 SPRY_Mosaic/v7.36 (Windows 16-bit) SPRY_package/v4.00
 5 PRODIGY-WB/2.1b
 5 Infoseek Robot 1.17
 5 ArchitextSpider
 4 Webinator-mass.thunderstone.com/1.21
```

```
4 Microsoft Internet Explorer/4.40.308 (Windows 95)
3 OTI_Robot/2.0  libwww/2.17
3 IBM WebExplorer /v1.0
2 Wobot/1.00
2 Lynx/2.5FM  libwww-FM/2.14
2 Lynx/2-4-2  libwww/2.14
2 InfoSeek Sidewinder
1 mytestlinks/0.3 libwww-perl/0.40
1 Lynx/2.3 BETA  libwww/2.14
  via proxy gateway  CERN-HTTPD/3.0 libwww/2.17
1 IWENG/1.2.000  via proxy gateway  CERN-HTTPD/3.0 libwww/2.17
1 HTMLgobble/2.3c
```

Does Anyone Really Care?

The reality check question on user agent information is: do you really care? While these particular statistics might be somewhat amusing, their usefulness isn't readily apparent to most system administrators or to marketing types, for that matter. Sure, there are going to be occasions when a web site is targeted toward a particular browser or operating system user; in that case, some useful analysis might be possible. But those cases are going to be the exception rather than the rule. For the most part, user agent information is truly useful for only two purposes:

- Identifying bugs, errors, or logging anomalies encountered or created by a specific browser or version of a browser.
- Identifying and eliminating the effect of web spiders and robots inflating and distorting traffic statistics.

More Problems with Log Files

As I stated at the beginning of this chapter, user agent information is probably best logged combined in the transfer log, since the agent log doesn't provide any way to tie entries back to a particular visit in the transfer log. There's no time stamp, no hostname—no way at all to merge the information, unless you choose to log referrer and agent data in the transfer log. So if you think you might want to know what browser a particular visitor uses, you have no choice but to use the combined log format.

The next question is: Why doesn't everyone just use the combined log format? Many system administrators do use the combined log format, especially those who don't use log analysis tools or services. But there are a couple of reasons why the combined format won't work for everyone. The most important reason is that you might not be able to use the log analysis software or service that would otherwise be your top choice. Several of these tools and services can only work with transfer logs in strict Common Log Format. But these are mostly low-end shareware products. Most of the better log analysis tools can easily import and use multiple log formats including several combined formats.

Also, using the combined format can waste potentially significant amounts of disk space with redundant information in every hit of a visit. For example, say that a single reader's visit to your web site generates a total of 100 hits on your web pages and graphics. That's 100 log entry lines in your transfer log. Each of those 100 entries has some useful and unique information about each hit. But there's only one piece of useful referrer information that applies to this reader's visit—the URL from which he or she came. And that's recorded in the very first hit that they generate on your web site. The referrer field for the rest of the hit entries will record the URL of the page on your own web site that readers follow to arrive at the new page—information you already have in the chronological sequence of hits.

Similarly, user agent information recorded in the transfer log will be redundant. During a visit to your web site, the reader will use only one web browser. If you combine the user agent data into the transfer log, the server will repeatedly log the name of the reader's browser on each line of the transfer log.

Given this, it makes sense that the original web server developers decided to put referrer and user agent information in another, separate file. The trouble is in how they did it. Even if servers write user agent data to a separate log file, they still do it for every hit. So in the scenario we drew above where a single visit results in 100 hits, web servers will dutifully write log entries to the agent log and referrer log—saving little, if any, disk space.

Here's another downside to the combined format: when you combine referrer and user agent data with the transfer log, the length of the log entry lines gets very long—making it very difficult to browse and view a log file. Nearly every line will wrap around, making it difficult to follow. The system administrator is forced to import the transfer log into a database management system or develop custom filters to screen out redundant information and make the data easier to read.

To date, the focus of server developers has been on features and performance; this is good. It's given us powerful servers at excellent prices. But hopefully sometime soon, their attention will turn to cleaning up the deficiencies in logging schemes. It can be much better than it is.

In the next chapter, we will bring together what we've learned about each of the server log files. We will discover every possible piece of information about a visit that is possible to determine, and we will look at some rudimentary ways to summarize and present this information in a format that can be followed by human eyes.

COLOR

PART TWO

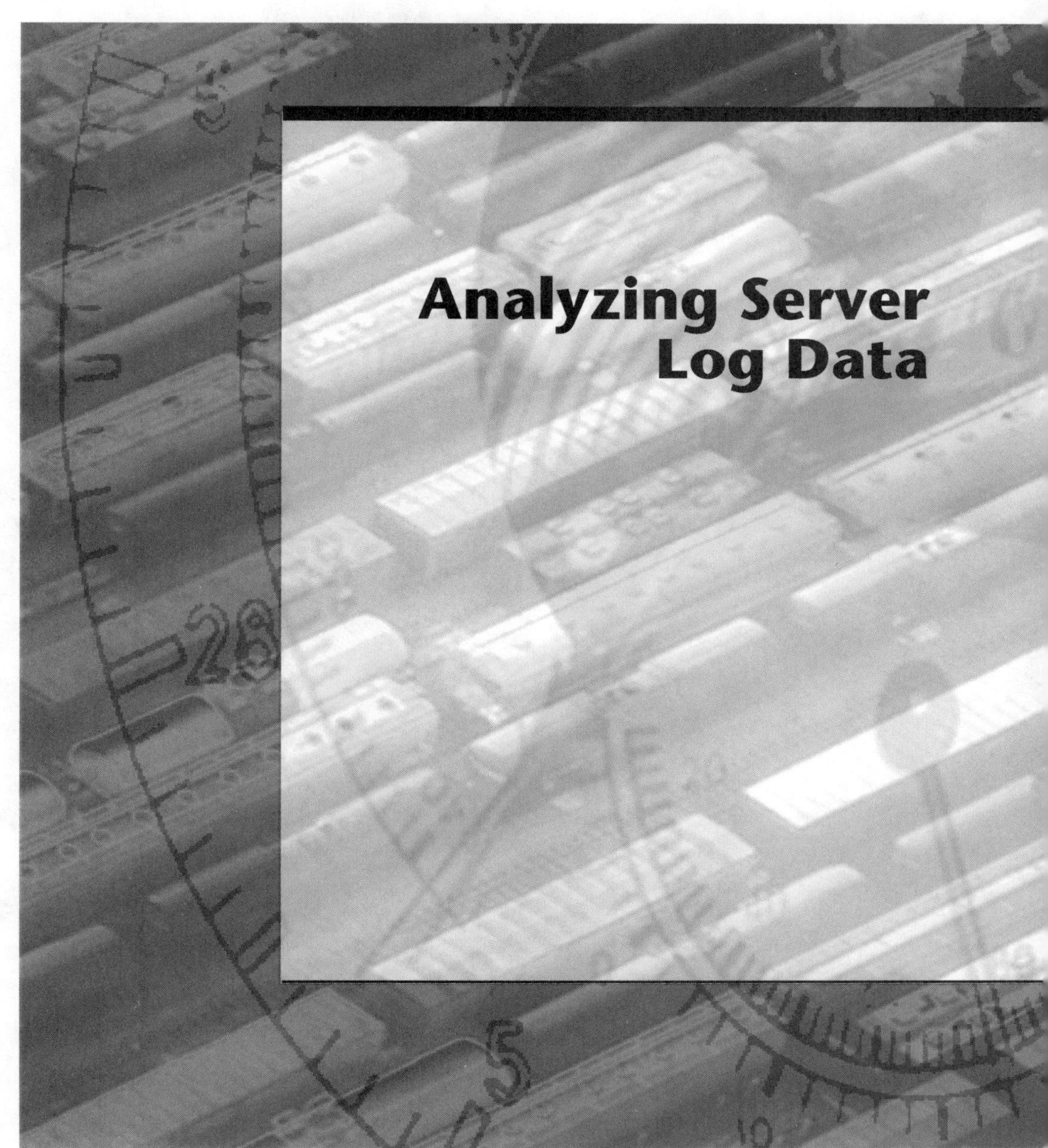

Analyzing Server Log Data

Chapter Six

The Visit—The Big Picture

In Part 1, you learned about the nitty-gritty details of web server log files. We looked at each log independently, studying each type of entry and each field. Now you should have a fairly solid understanding of the data that servers record. In Part 2, and particularly in this chapter, we're going to bring it all together to show you how to build a complete picture of a visit. We will examine, in great detail, one particular visit to a golf-related web site that we maintain.

This may well be the only time that you will ever examine a visit in such detail, at least as formally and methodically as we will here. With an active web site, it just isn't practical to go over every—or even an occasional—visit with the kind of fine-tooth comb we are going to use here. This will be a good exercise, though. By working through the details of this visit, you'll get an idea of the extent of information it's possible to extract from your server log files.

Working through this visit, peripheral subjects will come up as we notice interesting aspects of the visit. Some are important; others border on the trivial. But we will leave no stone unturned as we mine the log files for information.

Revisiting the Visit

When you key an address into your web browser's Location box or follow a hypertext link on a web page to go to a new site, the web browser goes through a sequence of events to contact the host; if it succeeds, the browser and server carry on a conversation. This conversation is called an *HTTP transaction,* and it consists of an HTTP request (made by the browser) and an HTTP response (made by the server).

It's important to understand that every hit (every distinct line, or entry) in a server's transfer log constitutes one of these HTTP transactions and is treated individually. As we work through this visit, it could be easy to come up with the impression that the server somehow knows that the collection of hits we are looking at is logically grouped together in what we're calling a visit. But the server doesn't have any really good way to know this.

For example, most people surfing the web today do so from their own single-user computer, such as a PC running Windows or a Macintosh. When these readers visit your web site, you can be reasonably sure that the tracks they leave in your log files relate to only them. But there are plenty of people around using multiuser computers, such as Unix systems, who are one of many users of that computer. The trouble is that there's no way for a web server to tell a single-user computer from a multiuser computer. A computer with a hostname like pc12.xyz.com could be a Windows 95 workstation with a single user, or it could be a Unix computer with 38 users concurrently online and browsing the web.

To make the assumption that hits from a single hostname come from the same person would be misleading. And if you make this assumption and three or four of those 38 users on a multiuser computer happen to visit your site at the same time, you'll end up with a hopelessly incomprehensible single "visit" from that host, which will distort any statistics you derive from your log data.

Before we begin our examination of this visit, there's one other caveat we should air. If you peruse your own transfer log and extract hits from a particular host the way we do here, it's very possible that the time stamps of the hits for graphics referenced in HTML pages won't be in as orderly or sequential a manner as they are here. On our golf site server, we use a run-time configuration option called a KeepAlive directive. The KeepAlive directive tells the server not to start up a new process for each HTTP request coming through an open connection.

For example, the top-level page of the golf site references some 14 objects besides the HTML document itself. Without the KeepAlive directive, the server may start up new processes to transfer each of these objects, effectively transferring them all simultaneously. This puts unnecessary load on the server and its bandwidth and makes it harder (if not impossible) to determine how long transfers are taking. By using the KeepAlive directive, you tell the server to service requests from the same host sequentially.

Getting Started

I used the Unix more command to page through our golf site's transfer log and noticed a somewhat wholesome visit from a host named pc12.xyz.com (actually, we've changed the hostname to protect the poor fellow—let's call him Bob—after all, he was probably on company time).

Next, we used the Unix grep command to pull every log entry for pc12.xyz.com out of the transfer log and redirect those log entries into another file I named pc12. Here's the command I used:

```
% grep pc12.xyz.com > pc12
% wc -l pc12
%    80 pc12
%
```

A quick line count (the wc -l command) shows that Bob registered 80 separate hits on our web site during his visit.

Figure 6-1 shows the contents of the pc12 file. However, I've removed the first three fields of each line (the hostname, rfc931 field, and the authuser field) so that you can see the important parts—the last four fields.

Since web servers log requests sequentially, there's no need to sort the log lines—they're already in chronological order. The grep command that we used to extract these lines from the transfer log also works sequentially through the file it operates on, so it doesn't affect the order. You can check this if you like by looking at the time stamps of each entry and verifying that each entry has a later time than the entry before it.

First, let's make a general observation about this collection of log entries. Notice that the time stamp of the first entry is the following:

```
[01/Aug/1996:15:16:19 -0700]
```

and the last time stamp is this:

```
[01/Aug/1996:15:26:46 -0700]
```

When Bob requested the first object from the golf site, the time was 15:16:19 (3:16:19 P.M.). When he requested his last object, the time was 15:26:46. So although Bob hit a lot of content while he was at the site, he did it all in about ten minutes.

Now, let's break down the entries into pieces and follow Bob's path through the web site. Here are the first five entries:

```
[01/Aug/1996:15:16:19 -0700] "GET /class/clssmisc/msgs27233.html HTTP/1.0" 302 -
[01/Aug/1996:15:16:20 -0700] "GET /class/clssmisc/msgs27233.html HTTP/1.0" 404 740
[01/Aug/1996:15:16:22 -0700] "GET /backgnd.gif HTTP/1.0" 200 3865
[01/Aug/1996:15:16:22 -0700] "GET /golfcrc2.gif HTTP/1.0" 200 5629
[01/Aug/1996:15:16:23 -0700] "GET /golfcirc.gif HTTP/1.0" 200 8966
```

Notice that Bob's first request was for a specific file (/class/clssmisc/msgs27233.html) instead of the usual starting point—simply the default file in the web site's root document directory, which would normally be /index.html, or just /. So now Bob's got us curious as to just how he came to try that URL first. Maybe he looked at the page before and bookmarked it, or maybe he followed a link on another web site.

However, we'll investigate this further in a minute, because besides that there are a couple more interesting things about Bob's arrival. His first request resulted in a status 302, which means the server redirected his request. The second request—which looks like it's for the very same file—got a status 404 (Not found) code. Let's see if we can sort this out.

```
[01/Aug/1996:15:16:19 -0700] "GET /class/clssmisc/msgs27233.html HTTP/1.0" 302 -
[01/Aug/1996:15:16:20 -0700] "GET /class/clssmisc/msgs27233.html HTTP/1.0" 404 740
[01/Aug/1996:15:16:22 -0700] "GET /backgnd.gif HTTP/1.0" 200 3865
[01/Aug/1996:15:16:22 -0700] "GET /golfcrc2.gif HTTP/1.0" 200 5629
[01/Aug/1996:15:16:23 -0700] "GET /golfcirc.gif HTTP/1.0" 200 8966
[01/Aug/1996:15:16:52 -0700] "GET / HTTP/1.0" 200 5047
[01/Aug/1996:15:16:54 -0700] "GET /art/backgnd.gif HTTP/1.0" 200 3865
[01/Aug/1996:15:16:59 -0700] "GET /golfcirc1.gif HTTP/1.0" 200 3177
[01/Aug/1996:15:16:59 -0700] "GET /header.gif HTTP/1.0" 200 4961
[01/Aug/1996:15:16:59 -0700] "GET /leader.gif HTTP/1.0" 200 1285
[01/Aug/1996:15:17:05 -0700] "GET /on_tour/pga/british/logo.gif HTTP/1.0" 200 2827
[01/Aug/1996:15:17:05 -0700] "GET /on_tour/pga/cvs/logo.gif HTTP/1.0" 200 3765
[01/Aug/1996:15:17:08 -0700] "GET /on_tour/pga/british/open.gif HTTP/1.0" 200 4445
[01/Aug/1996:15:17:14 -0700] "GET /circdir.gif HTTP/1.0" 200 3621
[01/Aug/1996:15:17:14 -0700] "GET /new.gif HTTP/1.0" 200 116
[01/Aug/1996:15:17:15 -0700] "GET /directory/circnet.gif HTTP/1.0" 200 3446
[01/Aug/1996:15:17:15 -0700] "GET /greendot.gif HTTP/1.0" 200 326
[01/Aug/1996:15:17:16 -0700] "GET /netscape.gif HTTP/1.0" 200 1135
[01/Aug/1996:15:17:18 -0700] "GET /Banners.class HTTP/1.0" 200 23912
[01/Aug/1996:15:17:26 -0700] "GET /BannersMsg.class HTTP/1.0" 200 1376
[01/Aug/1996:15:18:06 -0700] "GET /tips/qa2.html HTTP/1.0" 200 3485
[01/Aug/1996:15:18:08 -0700] "GET /adnet/odyssey2.gif HTTP/1.0" 200 7555
[01/Aug/1996:15:18:10 -0700] "GET /tips/backgnd.gif HTTP/1.0" 200 3831
[01/Aug/1996:15:18:12 -0700] "GET /tips/tips.gif HTTP/1.0" 200 6445
[01/Aug/1996:15:21:53 -0700] "GET /golfcrc2.map?167,13 HTTP/1.0" 302 -
[01/Aug/1996:15:21:54 -0700] "GET /tips/ HTTP/1.0" 200 1642
[01/Aug/1996:15:21:56 -0700] "GET /adnet/odyssey3.gif HTTP/1.0" 200 3205
[01/Aug/1996:15:21:56 -0700] "GET /tips/backgnd.gif HTTP/1.0" 200 3831
[01/Aug/1996:15:21:56 -0700] "GET /tips/tips.gif HTTP/1.0" 200 6445
[01/Aug/1996:15:21:57 -0700] "GET /tips/tips1.gif HTTP/1.0" 200 3701
[01/Aug/1996:15:22:49 -0700] "GET /golfcrc2.map?77,10 HTTP/1.0" 302 -
[01/Aug/1996:15:22:50 -0700] "GET /news/ HTTP/1.0" 200 2048
[01/Aug/1996:15:22:53 -0700] "GET /news/ad.html HTTP/1.0" 200 789
[01/Aug/1996:15:22:53 -0700] "GET /news/corner.html HTTP/1.0" 200 222
[01/Aug/1996:15:22:53 -0700] "GET /news/main.html HTTP/1.0" 200 680
[01/Aug/1996:15:22:53 -0700] "GET /news/menu.html HTTP/1.0" 200 894
[01/Aug/1996:15:22:54 -0700] "GET /on_tour/pga/british/logo.gif HTTP/1.0" 200 2827
[01/Aug/1996:15:22:55 -0700] "GET /leader.gif HTTP/1.0" 200 1285
[01/Aug/1996:15:22:56 -0700] "GET /news/AdvertisementPanel.class HTTP/1.0" 200 6624
[01/Aug/1996:15:22:56 -0700] "GET /news/news.gif HTTP/1.0" 200 5149
[01/Aug/1996:15:22:58 -0700] "GET /news/backgnd.gif HTTP/1.0" 200 3831
[01/Aug/1996:15:22:58 -0700] "GET /news/select1.gif HTTP/1.0" 200 1317
[01/Aug/1996:15:23:02 -0700] "GET /news/espn2.gif HTTP/1.0" 200 3187
[01/Aug/1996:15:23:03 -0700] "GET /news/select2.gif HTTP/1.0" 200 1209
[01/Aug/1996:15:23:04 -0700] "GET /news/select3.gif HTTP/1.0" 200 1284
[01/Aug/1996:15:23:04 -0700] "GET /news/select4.gif HTTP/1.0" 200 1337
[01/Aug/1996:15:23:06 -0700] "GET /ads/tid.jpg HTTP/1.0" 200 5533
[01/Aug/1996:15:23:06 -0700] "GET /ads/virtfair.gif HTTP/1.0" 200 8387
[01/Aug/1996:15:23:07 -0700] "GET /ads/ad1.gif HTTP/1.0" 200 8008
[01/Aug/1996:15:23:07 -0700] "GET /ads/golfbags.gif HTTP/1.0" 200 1893
[01/Aug/1996:15:23:15 -0700] "GET /news/greendot.gif HTTP/1.0" 200 326
[01/Aug/1996:15:23:15 -0700] "GET /news/pga.gif HTTP/1.0" 200 1208
[01/Aug/1996:15:23:17 -0700] "GET /news/spga.gif HTTP/1.0" 200 1237
[01/Aug/1996:15:23:18 -0700] "GET /news/lpga.gif HTTP/1.0" 200 1231
[01/Aug/1996:15:23:18 -0700] "GET /news/nike.gif HTTP/1.0" 200 1223
[01/Aug/1996:15:24:19 -0700] "GET /golfcrc2.map?330,9 HTTP/1.0" 302 -
[01/Aug/1996:15:24:20 -0700] "GET /proshop.html HTTP/1.0" 200 1908
[01/Aug/1996:15:24:22 -0700] "GET /adnet/odyssey1.gif HTTP/1.0" 200 4161
[01/Aug/1996:15:24:22 -0700] "GET /backgnd.gif HTTP/1.0" 200 3865
[01/Aug/1996:15:24:23 -0700] "GET /club.gif HTTP/1.0" 200 1387
[01/Aug/1996:15:24:27 -0700] "GET /bags1.gif HTTP/1.0" 200 1437
[01/Aug/1996:15:24:27 -0700] "GET /irons.gif HTTP/1.0" 200 1347
[01/Aug/1996:15:24:27 -0700] "GET /putters.gif HTTP/1.0" 200 1383
[01/Aug/1996:15:24:29 -0700] "GET /balls.gif HTTP/1.0" 200 1354
[01/Aug/1996:15:24:30 -0700] "GET /apparel.gif HTTP/1.0" 200 1403
[01/Aug/1996:15:24:30 -0700] "GET /trrypins.gif HTTP/1.0" 200 1475
[01/Aug/1996:15:24:31 -0700] "GET /newprods.gif HTTP/1.0" 200 1507
[01/Aug/1996:15:24:32 -0700] "GET /new.gif HTTP/1.0" 200 116
[01/Aug/1996:15:24:33 -0700] "GET /cards.gif HTTP/1.0" 200 5298
[01/Aug/1996:15:24:40 -0700] "GET /fan002.gif HTTP/1.0" 200 13218
[01/Aug/1996:15:25:02 -0700] "GET /proshop/balls.html HTTP/1.0" 200 525
[01/Aug/1996:15:25:04 -0700] "GET /proshop/dimple.gif HTTP/1.0" 200 380
[01/Aug/1996:15:25:04 -0700] "GET /proshop/greendot.gif HTTP/1.0" 200 326
[01/Aug/1996:15:25:20 -0700] "GET /proshop/titleist/titlball.html HTTP/1.0" 200 1244
[01/Aug/1996:15:25:21 -0700] "GET /proshop/titleist/white.gif HTTP/1.0" 200 43
[01/Aug/1996:15:25:25 -0700] "GET /proshop/titleist/titlball.gif HTTP/1.0" 200 13837
[01/Aug/1996:15:26:02 -0700] "GET /proshop/bridgestone/brdgball.html HTTP/1.0" 200 13
[01/Aug/1996:15:26:04 -0700] "GET /proshop/bridgestone/white.gif HTTP/1.0" 200 43
[01/Aug/1996:15:26:08 -0700] "GET /proshop/bridgestone/precept.gif HTTP/1.0" 200 1491
[01/Aug/1996:15:26:46 -0700] "GET /class/clssmisc/msgs27233.html HTTP/1.0" 404 740
```

Figure 6-1. *Transfer log hits from pc12.xyz.com (last four fields only)*

Originally, the domain name for our golf site was sdgolf.com. Later, Tom, the site's owner, changed it to golfcircuit.com, but by that time, it was already listed in a number of search engines and had been worked over regularly by numerous spiders and robots. Simply changing the domain name would undo much of what Tom had done to attract readers. Potential new readers who would try to follow links to sdgolf.com listed in a search engine and repeat readers who had bookmarked our site would think we had just fallen off the Net into the big bit bucket.

So, rather than simply changing our domain name, we added the new one and kept the original. We went through the site and changed all of the hypertext links to refer to the new domain. In addition, we configured our web server to redirect any requests for documents on the old domain to the new domain. This is a standard feature of nearly every web server. So now any incoming request for a document on the sdgolf.com domain would be automatically redirected to the golfcircuit.com domain. For example, requests for

```
http://www.sdgolf.com
```

or

```
http://www.sdgolf.com/foo/bar
```

would be automatically redirected to

```
http://www.golfcircuit.com
```

and

```
http://www.golfcircuit.com/foo/bar
```

In addition to configuring the server to redirect any requests to the new URL, we configured it to use the same log files for both sites. This is why we see both the redirect and a second (apparently) identical request. The first line is the redirect from the old domain, and the second is the request readdressed to the new domain.

Why did the second request fail with a status 404? And if it failed, what are those 740 bytes transferred as a result of the request? A quick look in the directory referenced reveals that the request failed because the file indeed doesn't exist. This directory holds classified ads that our readers place themselves through an HTML form and a CGI program. When they've sold their merchandise or found what they're looking for, they remove the ad themselves as well. Maybe Bob saw an ad for a killer set of clubs and decided to bookmark it until he could think of a way to justify them to his wife...and he simply took too long.

We don't need to speculate, though. Let's see if this file appears in the referrer log. When we grep the referrer log for msgs27233.html, we find this line:

```
http://www.excite.com/search.gw?search=GOLF,HANDICAP -> /class/clssmisc/msgs27233.html
```

Aha! This explains it. Bob entered some search criteria into a search engine and was presented with a page that had a link to an old classified ad on our site. Just to make sure, we can go to the same search engine and enter the same search criteria and see what comes up. Figure 6-2 shows the resulting page from the Excite search engine. Sure enough, our classified ad is listed (it's the second in the list).

Let's review what we know so far about Bob's visit. He found his way to the Excite search engine because he was interested in golf or golf handicapping. He performed a search on the words "golf" and "handicap," which presented him with a results page that had an entry for an old classified ad on our golf site. (It was so old, in fact, that it referenced the old domain name—and was long since deleted anyway.)

However, it looked interesting to Bob, so he clicked on the link. Our web server intercepted his request for a page on the old sdgolf.com domain and redirected it to golfcircuit.com. Then the server attempted to handle the redirected request and found that the file didn't exist anymore, so it issued a status 404 (Not found) code, and—oops—why did the server transfer 740 bytes for a nonexistent page?

Another one of the run-time options of many web servers allows you to specify a special error document to send to clients in the case of any particular error that you specify. We've done this on the golfcircuit site. Using our server's ErrorDocument directive, we've directed the server to send the file /notfound.html in the event of any status 404 error conditions. This allows you to put a pretty face on an ugly situation—or at least a bag over an ugly raised head. Figure 6-3 shows our error document for the status 404.

Our /notfound.html file contains exactly 740 bytes of HTML code. It also references three inline images—a background image, the title image, and the imagemap bar at the bottom of the page. This explains the remaining three hits in the first five lines we're looking at.

Presented with our error document page, Bob clicked the "Click here to continue" link, which brought him to our top-level page (the /index.html that

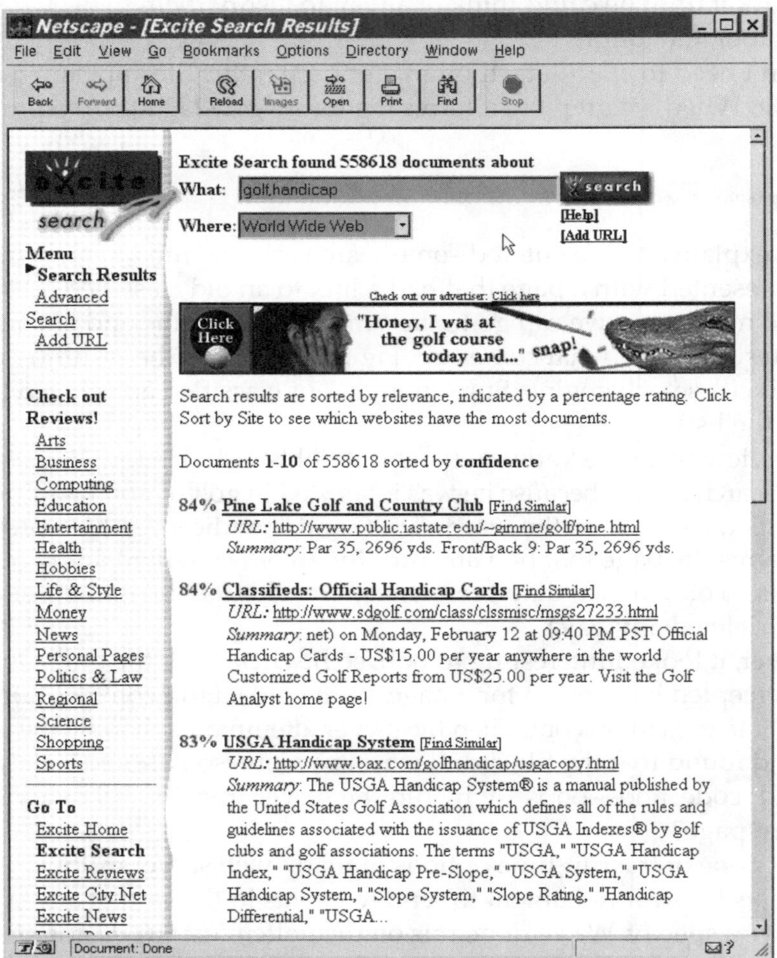

Figure 6-2. *Search results from the Excite search engine*

is shown simply as /). Now let's focus on this hit and all of the hits following, up to the point where another .html file is requested. Here are the hits:

```
[01/Aug/1996:15:16:52 -0700] "GET / HTTP/1.0" 200 5047
[01/Aug/1996:15:16:54 -0700] "GET /art/backgnd.gif HTTP/1.0" 200 3865
[01/Aug/1996:15:16:59 -0700] "GET /golfcircl.gif HTTP/1.0" 200 3177
[01/Aug/1996:15:16:59 -0700] "GET /header.gif HTTP/1.0" 200 4961
[01/Aug/1996:15:16:59 -0700] "GET /leader.gif HTTP/1.0" 200 1285
[01/Aug/1996:15:17:05 -0700] "GET /on_tour/pga/british/logo.gif HTTP/1.0" 200 2827
```

```
[01/Aug/1996:15:17:05 -0700] "GET /on_tour/pga/cvs/logo.gif HTTP/1.0" 200 3765
[01/Aug/1996:15:17:08 -0700] "GET /on_tour/pga/british/open.gif HTTP/1.0" 200 4445
[01/Aug/1996:15:17:14 -0700] "GET /circdir.gif HTTP/1.0" 200 3621
[01/Aug/1996:15:17:14 -0700] "GET /new.gif HTTP/1.0" 200 116
[01/Aug/1996:15:17:15 -0700] "GET /directory/circnet.gif HTTP/1.0" 200 3446
[01/Aug/1996:15:17:15 -0700] "GET /greendot.gif HTTP/1.0" 200 326
[01/Aug/1996:15:17:16 -0700] "GET /netscape.gif HTTP/1.0" 200 1135
[01/Aug/1996:15:17:18 -0700] "GET /Banners.class HTTP/1.0" 200 23912
[01/Aug/1996:15:17:26 -0700] "GET /BannersMsg.class HTTP/1.0" 200 1376
```

All 15 of these hits are the result of Bob loading our top-level page. The page itself (the .html file) contains 5,047 bytes of HTML code and references a number of inline images and even a couple of Java applets.

There don't seem to be any errors or mysteries to decode here, but there are some interesting conclusions you can draw from this segment of the transfer log. First, notice that every hit got a status 200 (OK). That's good—there are no broken inline links or missing image files. But if you didn't know it already,

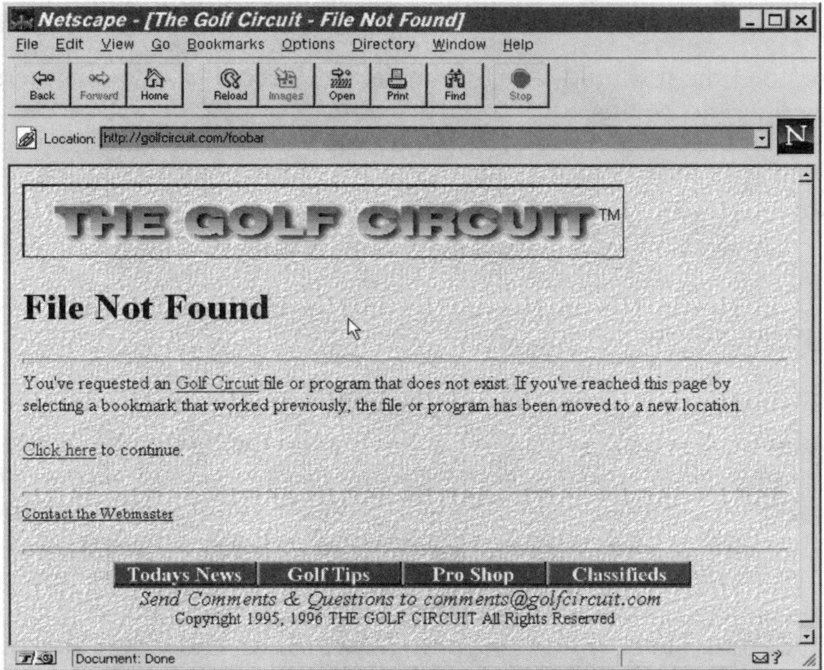

Figure 6-3. *An error document specified by an ErrorDocument directive*

you can see that this page—with all of its graphics and Java applets—is a bit on the porky side.

If you add up the bytes transferred for each of these hits, you'll find the grand total of the page to be 63,304 bytes. If you take the difference in the first and last time stamps, you can see that it took Bob 34 seconds to complete his download of the complete page (actually, that doesn't include the last Java applet, since time stamps indicate the time of the request rather than the time the server completes its transfer of the file).

If we really wanted to get nitpicky about estimating Bob's transfer time, we could do something like this: in the 34 seconds prior to beginning the last transfer (the /BannersMsg.class object), the server successfully transferred 61,928 bytes. Or, to put it mathematically:

61,928 bytes

--------------- = 1,821 bytes/second

34 seconds

Therefore, the time it should take to transfer the last object (the /Banners Msg. class object) would be this:

1,376 bytes

--------------- = .76 seconds

1,821 bytes/second

You can estimate that Bob's total transfer time for the golf site's top-level page at around 35 seconds. Does this mean that it takes everyone 35 seconds to load the page? Not at all. Every system has a different kind of connection to the Internet. Internet service providers have connections to the Internet of varying throughput, and every dial-up user has individual limitations.

Let's see, though, if we can figure out where Bob's throughput bottleneck is. Let's guess that he's using a modem for his Internet connection and compare the time it took him to download the page (of which we know the data volume) to the throughput capabilities of popular modems (which we also know).

Modem speeds are expressed in bits per second. The first thing we have to do is convert some popular modem speeds to bytes per second. Perhaps the most popular modem today is the 28,800 bps (or 28.8 kbps) modem. Since there are 8 bits in a byte, 28,800 bits per second is equivalent to 3,600 bytes per second. Of course, a 28.8 modem only really gets this throughput when it's doing full compression and is working with optimally compressible data.

At 3,600 bytes per second, our top-level page (of 63,304 bytes) should download in 18 seconds. Just about half the time it took Bob—or about what it should take for a 14.4 kbps modem. So it's a fairly good bet that Bob's bottleneck is a modem (it's probably his modem) and that it's either a 28.8 kbps modem not doing compression or it's a 14.4 kbps modem that is successfully doing compression.

Can you take any of this to the bank? You'd better not. First of all, no 28,800 bps modem ever really gets a 3,600 byte per second throughput. There are the framing bits (start and stop bits) of the modem protocol, and there are latencies and overhead in TCP/IP packeting that are very hard to predict. In short, there are so many variables that can come into play with the Internet that these particular statistics are no better than the results of a random telephone poll. But it is fun to do, and it may occasionally even be close to the reality of the situation.

Let's move on to the next page that Bob loads.

```
[01/Aug/1996:15:18:06 -0700] "GET /tips/qa2.html HTTP/1.0" 200 3485
[01/Aug/1996:15:18:08 -0700] "GET /adnet/odyssey2.gif HTTP/1.0" 200 7555
[01/Aug/1996:15:18:10 -0700] "GET /tips/backgnd.gif HTTP/1.0" 200 3831
[01/Aug/1996:15:18:12 -0700] "GET /tips/tips.gif HTTP/1.0" 200 6445
[01/Aug/1996:15:21:53 -0700] "GET /golfcrc2.map?167,13 HTTP/1.0" 302 -
```

Bob clicked on a button on the top-level page to move to this document (/tips/qa2.html), a question-and-answer (Q&A) tips page. Let's look first at the time stamp of the hit for the page itself. Bob clicked on a link to this page at 3:18:06 P.M. (15:18:06). The time stamp of the last hit on an HTML page (the / page) was 15:16:52. When we looked at the detail of Bob's last page view, we looked at how long it took him to download the page, and we speculated about what type of Internet connection he had. But not until now (as we look at the next hit on an HTML document) can we calculate how much time Bob actually spent looking at that first page.

The difference in time between the two time stamps (15:16:52 and 15:18:06) is exactly one minute and fourteen seconds (1:14). Does this mean that Bob spent exactly that amount of time reading the top-level page before following the link to the tips page? Not exactly. Remember that it took about 35 seconds for Bob's browser to load the top-level page. Assuming the text of the top-level page was readable nearly immediately (it took only two seconds to load the actual HTML page itself), it's possible that Bob was reading parts of the page as the graphics were loading. He probably didn't get a great deal out of the page until all of the graphical buttons and imagemaps completed loading as well, so Bob's real viewing time before clicking on the Tips button was probably closer to 39 seconds.

Unfortunately, most log analysis tools and services don't take this into account. They consider the time viewing a page to be merely the difference between the time stamps of the two consecutive HTML hits.

A few additional observations you can make about this page view: the page and its graphics are considerably smaller than the top-level page; and the HTML file itself and its three inline graphics come in at 21,316 bytes—or about a third of the volume of the top-level page.

Using the transfer rate (1,821 bytes/second) that we calculated on the top-level page view, it would take about 3.5 seconds to download the last graphic for the page /tips/tips.gif (which contains 6,445 bytes). Therefore, it took Bob's browser about 9.5 seconds to download this page and its graphics—the difference between the first time stamp (15:18:06) and the time stamp of the last graphic to load on the page (15:18:12) plus 3.5 seconds.

The last hit in this sequence is on an imagemap. There are two types of imagemaps, known as server-side imagemaps and client-side imagemaps. With a client-side imagemap, the code for the imagemap (and the URLs that it leads to) are built into the HTML of a web page. When you download and view a page with a client-side imagemap, your browser knows how to handle a click on the imagemap, because it's already got it in memory.

A server-side imagemap, on the other hand, is handled at the server's end. When you click on a defined area of a server-side imagemap, your browser must send the coordinates of your mouse cursor (over the imagemap image) back to the server for it to handle. The server compares the mouse coordinates to the coordinates specified in a map file—a file that typically has a filename extension of .map—and decides where to send your browser. Mechanically, the mechanism that servers use to direct your browser to the target URL is the internal redirect.

When a reader clicks on an area of a server-side imagemap, the browser issues a "request" for the imagemap map file itself, passing the coordinates of the mouse as an argument. Servers are written to work with this behavior—you could say they "know" to parse the map file, find the description of the area in which the click was made, and redirect the reader's browser to the URL associated with that particular area. This results in a status code 302 (Redirect) for the hit generated on the map file.

One last item of interest about this sequence of hits before we move on: you can clearly see that the reader pondered this page for nearly four minutes before clicking on the imagemap to move on.

Here is the next sequence of hits:

```
[01/Aug/1996:15:21:54 -0700] "GET /tips/ HTTP/1.0" 200 1642
[01/Aug/1996:15:21:56 -0700] "GET /adnet/odyssey3.gif HTTP/1.0" 200 3205
[01/Aug/1996:15:21:56 -0700] "GET /tips/backgnd.gif HTTP/1.0" 200 3831
[01/Aug/1996:15:21:56 -0700] "GET /tips/tips.gif HTTP/1.0" 200 6445
[01/Aug/1996:15:21:57 -0700] "GET /tips/tips1.gif HTTP/1.0" 200 3701
[01/Aug/1996:15:22:49 -0700] "GET /golfcrc2.map?77,10 HTTP/1.0" 302 -
```

After reading the Q&A page of the Tips section, our reader Bob decided to take a look at the general Tips page. He did this by clicking on the Tips section of our standard imagemap bar, which is at the bottom of each page.

This page includes four graphics, all of which are relatively small. The page together with its graphics took around five seconds to load (15:21:54 to 15:21:57)—about three seconds plus a couple more to load the last graphic of 3,701 bytes. Assuming that's about right, Bob's browser finished loading this page at approximately 15:21:59, and he clicked on the imagemap bar at 15:22:49, so he spent about 50 seconds on this page before moving on.

```
[01/Aug/1996:15:22:50 -0700] "GET /news/ HTTP/1.0" 200 2048
[01/Aug/1996:15:22:53 -0700] "GET /news/ad.html HTTP/1.0" 200 789
[01/Aug/1996:15:22:53 -0700] "GET /news/corner.html HTTP/1.0" 200 222
[01/Aug/1996:15:22:53 -0700] "GET /news/main.html HTTP/1.0" 200 680
[01/Aug/1996:15:22:53 -0700] "GET /news/menu.html HTTP/1.0" 200 894
[01/Aug/1996:15:22:54 -0700] "GET /on_tour/pga/british/logo.gif HTTP/1.0" 200 2827
[01/Aug/1996:15:22:55 -0700] "GET /leader.gif HTTP/1.0" 200 1285
[01/Aug/1996:15:22:56 -0700] "GET /news/AdvertisementPanel.class HTTP/1.0" 200 6624
[01/Aug/1996:15:22:56 -0700] "GET /news/news.gif HTTP/1.0" 200 5149
[01/Aug/1996:15:22:58 -0700] "GET /news/backgnd.gif HTTP/1.0" 200 3831
[01/Aug/1996:15:22:58 -0700] "GET /news/select1.gif HTTP/1.0" 200 1317
[01/Aug/1996:15:23:02 -0700] "GET /news/espn2.gif HTTP/1.0" 200 3187
[01/Aug/1996:15:23:03 -0700] "GET /news/select2.gif HTTP/1.0" 200 1209
[01/Aug/1996:15:23:04 -0700] "GET /news/select3.gif HTTP/1.0" 200 1284
[01/Aug/1996:15:23:04 -0700] "GET /news/select4.gif HTTP/1.0" 200 1337
[01/Aug/1996:15:23:06 -0700] "GET /ads/tid.jpg HTTP/1.0" 200 5533
[01/Aug/1996:15:23:06 -0700] "GET /ads/virtfair.gif HTTP/1.0" 200 8387
[01/Aug/1996:15:23:07 -0700] "GET /ads/ad1.gif HTTP/1.0" 200 8008
[01/Aug/1996:15:23:07 -0700] "GET /ads/golfbags.gif HTTP/1.0" 200 1893
[01/Aug/1996:15:23:15 -0700] "GET /news/greendot.gif HTTP/1.0" 200 326
[01/Aug/1996:15:23:15 -0700] "GET /news/pga.gif HTTP/1.0" 200 1208
[01/Aug/1996:15:23:17 -0700] "GET /news/spga.gif HTTP/1.0" 200 1237
[01/Aug/1996:15:23:18 -0700] "GET /news/lpga.gif HTTP/1.0" 200 1231
[01/Aug/1996:15:23:18 -0700] "GET /news/nike.gif HTTP/1.0" 200 1223
[01/Aug/1996:15:24:19 -0700] "GET /golfcrc2.map?330,9 HTTP/1.0" 302 -
```

This next sequence of hits has a bunch of activity. Bob decided to check out our News page. This page employs frames, one of Netscape's enhancements to

HTML. The default page here (/news/index.html) defines four separate framesets loading four different HTML pages: ad.html, corner.html, main.html, and menu.html. Between them, this page (or collection of pages, depending on how you like to look at framed pages) displays 18 discrete images, plus a Java applet.

It took Bob's browser about 30 seconds to pull the page down, and he looked it over for about a minute after it finished downloading.

```
[01/Aug/1996:15:24:20 -0700] "GET /proshop.html HTTP/1.0" 200 1908
[01/Aug/1996:15:24:22 -0700] "GET /adnet/odyssey1.gif HTTP/1.0" 200 4161
[01/Aug/1996:15:24:22 -0700] "GET /backgnd.gif HTTP/1.0" 200 3865
[01/Aug/1996:15:24:23 -0700] "GET /club.gif HTTP/1.0" 200 1387
[01/Aug/1996:15:24:27 -0700] "GET /bags1.gif HTTP/1.0" 200 1437
[01/Aug/1996:15:24:27 -0700] "GET /irons.gif HTTP/1.0" 200 1347
[01/Aug/1996:15:24:27 -0700] "GET /putters.gif HTTP/1.0" 200 1383
[01/Aug/1996:15:24:29 -0700] "GET /balls.gif HTTP/1.0" 200 1354
[01/Aug/1996:15:24:30 -0700] "GET /apparel.gif HTTP/1.0" 200 1403
[01/Aug/1996:15:24:30 -0700] "GET /trrypins.gif HTTP/1.0" 200 1475
[01/Aug/1996:15:24:31 -0700] "GET /newprods.gif HTTP/1.0" 200 1507
[01/Aug/1996:15:24:32 -0700] "GET /new.gif HTTP/1.0" 200 116
[01/Aug/1996:15:24:33 -0700] "GET /cards.gif HTTP/1.0" 200 5298
[01/Aug/1996:15:24:40 -0700] "GET /fan002.gif HTTP/1.0" 200 13218
```

Next, Bob stopped into our pro shop. There's not a lot to note about this page view. It has about a dozen graphics, and it took Bob about 20 seconds to load the objects up to the last line. The last object (fan002.gif) is somewhat larger than the rest at 13,218 bytes; it probably took another seven or so seconds to load that image. After viewing this page for about 15 seconds, Bob followed a link to a page sporting golf balls.

```
[01/Aug/1996:15:25:02 -0700] "GET /proshop/balls.html HTTP/1.0" 200 525
[01/Aug/1996:15:25:04 -0700] "GET /proshop/dimple.gif HTTP/1.0" 200 380
[01/Aug/1996:15:25:04 -0700] "GET /proshop/greendot.gif HTTP/1.0" 200 326
```

This is a fairly small page, with only a couple of graphics. In fact, it has only a few links that lead to separate pages for each of the several golf ball manufacturers. After about 16 seconds, Bob clicked on the link for Titleist brand balls.

```
[01/Aug/1996:15:25:20 -0700] "GET /proshop/titleist/titlball.html HTTP/1.0" 200 1244
[01/Aug/1996:15:25:21 -0700] "GET /proshop/titleist/white.gif HTTP/1.0" 200 43
[01/Aug/1996:15:25:25 -0700] "GET /proshop/titleist/titlball.gif HTTP/1.0" 200 13837
```

The page and the first graphic image loaded in about four seconds, and the last image took another seven or eight seconds to load.

```
[01/Aug/1996:15:26:02 -0700] "GET /proshop/bridgestone/brdgball.html HTTP/1.0" 200 1320
[01/Aug/1996:15:26:04 -0700] "GET /proshop/bridgestone/white.gif HTTP/1.0" 200 43
```

```
[01/Aug/1996:15:26:08 -0700] "GET /proshop/bridgestone/precept.gif HTTP/1.0" 200 14911
[01/Aug/1996:15:26:46 -0700] "GET /class/clssmisc/msgs27233.html HTTP/1.0" 404 740
```

Next, Bob backtracked by clicking the back arrow on his web browser and then followed a link for another ball manufacturer. There's actually no way to know this firsthand. But because there isn't a link on the Titleist page that leads to the Bridgestone page, we know that's what he must have done.

This is actually a fairly significant shortcoming of the current HTML protocol—at least as far as keeping stats is concerned. It's important to remember that the primary objectives of developers of the HTTP protocol, web servers, and web browsers are features and performance. With the protocol as it is today, there's no facility for recording a second view on a page that is in the memory cache of the reader's web browser. When a reader uses the back arrow on the browser to return to a page viewed previously, it's completely invisible to the server.

How important is it? If you don't give equal weight to accurate statistics as a design objective, you may not see it as important at all. However, don't be under the impression that page view durations can be accurate. Most of the log analysis tools and services available today do calculate average page view time, but they can't possibly be accurate as long as a step backward is an unrecorded event.

After Bob's visit to the pro shop, he tried to pull up the classified ad once again. Again, he got an error code because it's long since gone, and Bob was off on his way.

The Visit in Summary

For the most part, Bob's visit to our golf site was common. He found our site by using a search engine to search for an unrelated topic: golf and handicaps. The search engine produced a list of links, one of which led to our site, although to a page that was no longer there. Once he encountered our error document, though, he decided to take a look at the site anyway. Besides the error document, Bob looked at eight pages, and he spent about ten minutes at the site. Let's review his steps through the site.

Time	Duration	Page Viewed
15:16:19	33 sec	/notfound.html - error document
15:16:52	1 min 14 sec	/index.html - the top-level page
15:18:06	3 min 48 sec	/tips/qa2.html - the tips/Q&A page
15:21:54	56 sec	/tips/index.html - the main tips page

Time	Duration	Page Viewed
15:22:50	1 min 30 sec	/news/index.html - the news page
15:24:20	42 sec	/proshop.html - the pro shop
15:25:02	18 sec	/proshop/balls.html - the balls page
15:25:20	42 sec	/proshop/titleist/titlball.html - Titleist balls
15:26:02	44 sec	/proshop/titleist/bridgestone.html - Bridgestone balls
15:26:46	unknown	/notfound.html - error document

The duration column in this table is the absolute time between requests. It doesn't take into account the time it took Bob to download each page. So, although these numbers aren't really true, they are easy to compute and they're what most log analysis tools and services use.

We did check the error and agent logs for information on Bob's visit to our web site. With the exception of the 404 (Not found) errors that Bob encountered twice, he generated no error conditions. This isn't unusual, especially since we know that all of the links in our site are good. The only other error condition we would not have been surprised to see would be transfer cancellations. If Bob had grown impatient waiting for a page to load all of its graphic images, he might have bailed out by clicking the back arrow or clicking on a link he could already see on the new page. But he didn't.

As for the user agent log, since we are using the Common Log Format with separate log files for referrer and agent information, we have no way of knowing which lines in the referrer log relate to Bob. We could remedy this by using a combined, or extended, format for the transfer log, but we don't really care what browser Bob is using at this point.

In Chapter 7, we're going to summarize much of what you've learned in this chapter. We will also consider how to accumulate and study visits in aggregate. Since you'll seldom have the time to analyze individual visits—and because most log analysis tools produce summary statistics—we'll look at the science of summary statistics.

Chapter Seven

Tracking the Elusive Visit

In the last chapter, we went through a single visit in great detail. We looked at every hit, analyzing its cause and meaning. We broke down the entire collection of hits into logical segments to analyze them separately as a smaller collection of a half-dozen or so events (page views). Once we understood each of those events completely, we grouped them back together to paint a high-level picture of what actually transpired with that particular visit.

However, as we also mentioned in that chapter, you will seldom take the time to examine a visit in that much detail. After all, once you do, you only have information about a single visit. What is more important is how you can summarize information about many (or all) visits and present that information in an informative and useful way.

The biggest problem we encounter when we try to summarize visits isn't how to present the resulting information—it's how to acquire that information in the first place. In this chapter, we will look at the very mechanics of tracking visits and visitors. This will lead us through some peripheral subjects such as the anatomy of HTTP requests and responses. Along the way, we'll also look at how tracking visits and visitors can help your analysis of your site's traffic.

Why Visits Are Difficult to Track

In Chapter 6, before we began our examination of the visit, I explained how I selected those 80 hits out of the transfer log: I paged through the transfer log looking for an interesting visit to examine. I found one—but I made a fundamental assumption about what I found.

I found a collection of hits that originated from the same host computer (pc12.xyz.com), and they all happened within about a ten-minute time frame. Now, it's a pretty good bet that all of these hits were created by the same person, especially since we tracked the hits from the first to the last, and they all flowed logically and chronologically. So, in this case, my assumption was probably a good one. But we have to think about the implications of extending that same assumption to all such collections of hits.

Let's put ourselves in the shoes of a software developer. We're developing log analysis software, and we want, as a primary objective, to be able to track and summarize *visits* as opposed to merely *hits* on the transfer log. To begin, let's code our software to make the same assumptions I made earlier: that hits originating from the same host within a reasonable amount of time constitute a visit; when we haven't received any additional hits from a host after 30 minutes, we will assume the reader has left and the visit has concluded after the last hit. Making these assumptions, under what circumstances would statistics that we generate be wrong or misleading?

First, let's consider multiuser computer systems. Even though single-user Windows and Macintosh computers outnumber multiuser systems, there are still plenty of people who primarily use multiuser computers like Unix systems. Some Unix users use character-based (nongraphical) terminals and work with the computer from a command line; some of these folks actually enjoy and prefer to surf the web with a character-based web browser such as Lynx. Any fairly loaded multiuser Unix system could have a couple dozen, or even more, of these users. Since they all share the same "host," they all drop the same hostname in the web server log files of the sites they visit. If they visit the same site at the same time, there's no way to distinguish between their log entries. If they visit the same web sites on different days, there's no way to tell whether the same person was visiting multiple times or different users were each visiting only once.

In addition, there's another type of multiuser Unix system called an X Server (actually, there's probably not much that's different about the system itself, except that it runs a program called an X Server). The Unix equivalent to the Windows and Macintosh graphical environments is called X Window (or simply X). A Unix computer running X Window has a graphical interface much like that of Windows 95 or a Mac.

Like many of the Unix equivalents to PC and Mac features, X Window enjoys decades of development and refinement. Unlike Windows systems, a single Unix computer running an X Server can manage graphic displays for several (even many) not-so-smart terminals called X terminals. They're by no means dumb terminals—they have memory, graphic displays, mouses, and high-speed network connections—but they're not quite full-fledged computers. X terminals are actually quite common in corporate environments, especially in technology-focused companies. X terminal users running Netscape or other web browsers are all using the same host computer (the system running the X Server).

As you can see, even though they are outnumbered by PCs and Macs, multiuser computer systems are used by enough people to throw your traffic statistics well out of kilter.

How about the guy perusing your web site who suddenly gets an urge for a sandwich and to catch the last inning of a ball game? He may come back in 45 minutes and resume his visit to your web site, but your software will have declared his visit terminated after 30 minutes of inactivity. When he clicks on a link that is still being displayed on his monitor, your software will count it as the beginning of a whole new visit.

Perhaps worse than distorting the number of visits, your software will record this new visit as beginning in an odd location, proceeding on an erratic course through the site because much of the site will still be in the browser's memory cache and not registering any new hits at all. This will definitely throw a monkey wrench into your statistics, affecting things like the average number of pages viewed per visit and the average length of time spent viewing pages.

This example assumes that the reader has left his Internet connection up while he went for lunch or he has a fixed hostname and IP address. What if he's one of the many people who are assigned their IP addresses dynamically when they initiate their connection to their service provider? Odds are short that he will get a completely different IP address when he logs back in. Even if he takes only 10 minutes to eat the sandwich and returns to the computer immediately, when he reconnects and clicks on that link in his web browser, he'll have a different IP address and will be counted as a completely new visitor.

Almost certainly there are other reasons why our earlier assumptions about hostnames and time frames would fail us. Let's shift gears for a moment and think about what we would do with the additional visit information if we could get it and if it were reliable. What more would we be able to know or do with this information?

Why Visits Are Important

You've probably sensed by now that knowing about individual hits isn't enough, although you may not yet fully understand why we (and the entire web community) are trying to find a way to distill hit information down to visits, perhaps even visitors. And right about now you're probably thinking that you're about to find out the reason—but we're going to keep you in suspense just a little longer. Before we get into why everyone is trying to transition the perspective of log analysis from hits to visits, let's look at some of the statistics that log analysis packages produce today.

Generally, you can say that statistics break down into two categories:

- Statistics based on raw hits
- Statistics derived from the number of actual visits

Let's look first at the statistics you can derive from raw transfer log hits whether or not you have a reliable way to count the actual number of visits. These statistics include the following:

- Total number of requests
- Total bytes transferred
- Number of requests for each page, graphic, and file
- Top (most requested) documents
- Top (most requested) downloaded files
- Top submitted forms and scripts
- Top documents by directory
- Average number of requests per day or hour
- Average bytes transferred by hour
- Average number of hits on weekdays
- Average number of hits on weekend days

These are good statistics, but they leave a lot to be desired. We know that each hit has information about the requesting host—whether by pointing out the general assumptions we can make about the "type" of organization (from the host domain type) or by looking up actual company names and geographic information from a database. What good is any of this as long as the basis for the statistics are hits?

It makes no sense at all to derive and present statistics that represent, for example, the percentage of hits from Britain or the percentage of hits from a particular company. The number may actually be real, but what does it mean? You cannot reliably count visits, and you can't even come up with an average number of hits per visit to divide into the number of hits from Britain in order to estimate the number of visits from Britain.

Tracking visits is vitally important for a deeper analysis of your web site. When you can reliably track visits, the additional statistics you can derive are some of the most informative and interesting. A few of these are

- Number of visits
- Average number of requests per visit
- Average duration of a visit
- Visits from organizations (most active organizations)

- Visits by organization type
- Visits from countries (most active countries)
- Top visit entry pages
- Top page durations
- Top exit pages (last pages)
- Average number of visits per day/hour
- Geographic regions
- Top cities
- Top referring organizations
- Top referring URLs
- Top browsers
- User operating systems
- Average number of users on weekdays
- Average number of users on weekends
- Most active day of the week
- Most active day ever
- Number of hits on the most active day (with the most visits)

These statistics are compelling, and they can only take on any real meaning when they're based on actual visits as opposed to hits. This is why it's vitally important to track visits.

Tracking the Visit

Let's look at what some of the commercial log analysis developers are doing to fine-tune the problem of identifying visits and what they do with that information once they've come up with it. Let's use Intersé's Market Focus analysis package as an example. At the bottom of the analysis page that Market Focus produces, Intersé provides the definitions of terms it uses throughout the analysis. Intersé defines the term "visit" as "a series of consecutive requests from a user to an Internet site." It further clarifies that if a user doesn't make another request for 30 minutes, the previous collection of hits is considered a complete visit (sound familiar?).

Intersé's definition relies heavily on the term "user." So to understand the term "visit" as Intersé uses it, you must understand the term "user," and Intersé offers a somewhat more slippery definition for this term. Intersé explains that their software goes through a process of trying to identify visitors (users) uniquely. What this actually means is that it attempts to make an identification on two grounds before resorting to using the hostname and a time out period. In order of preference, Intersé's (and other developers') algorithms look for:

- Cookies
- Registered usernames
- Hostnames

If the first two methods fail to provide any solid way to track a visitor through a site, then the software has no option but to use the simple hostname field and the time out period methodology we described earlier.

Method two (registered usernames) refers to the authuser field in the transfer log that we discussed in detail in Chapter 2. The trouble with relying on the authuser field for identification is that in order to have this data, you have to require that visitors to your site register the first time they visit your site and then log in with a user ID and password thereafter. Virtually nobody requires this for general access to a web site. One of the main attractions about the web is that it allows people to explore and experience the wealth of information it contains almost completely anonymously. People enjoy the ease with which they can browse and window-shop, and they definitely do not enjoy having to register to be granted access.

Sites that do require registration almost always exist to serve a special purpose, for example, to allow only registered owners of a software package to obtain patches or upgrades, to disseminate valuable information only to paid members of an organization such as a professional association, or to allow customers to view reports created only for a particular customer—the way that NetCount (a log analysis service company) provides its reports to its customers.

Once visitors enter their username and password to access a secure part of the web site, you've got a lock on them. Every successive hit they generate will include their username in the authuser field of your transfer log. Of course, your log analysis software needs to be smart enough to go back and associate the hits the reader generated before logging into the secure area as well.

Even if web sites do have secure areas that are frequented by registered users, generally, the majority of the web site is not secured, and the majority of visitors will not be registered users. So the web site administrator will only be able to track a minority of visitors this way.

This brings us to the first method in the list for identifying and tracking unique visitors: cookies. Before we explain cookies, we have to take a few minutes to describe the anatomy of HTTP requests and responses. Once you have an idea of how HTTP transactions take place, we will introduce you to cookies.

The Anatomy of HTTP Requests and Responses

When a web browser gets ready to initiate a request to a web server, it first takes some information about itself, the request that the reader made by typing in a URL or clicking on a link, and the computer it's running on. The browser rolls up all of this information into a nice little bundle called an HTTP request. Think of an HTTP request as an electronic version of a typewritten memo that might read something like this:

```
To:  www.bluebikes.com
From:  win22.xyz.com
Subject:  Requesting a page
Browser:  Mozilla/3.2 (Windows 95)
Referrer: http://win22.xyz.com/coollinks.html
Request:  GET /aboutus.html HTTP/1.0

Dear Mr. Web Server,
Kindly send me the HTML object referenced above.
Thank you for your cooperation.
```

Okay, this might be a bit simplistic, but it conveys the purpose and even some of the structure of an actual HTTP request. Let's call the first six lines of this message the message *headers*. Notice there's no information at all in the body of this message that isn't spelled out even more explicitly in the message headers. So, since there really is no equivalent to the message body in a real HTTP request, let's get rid of that part. Also, there's really no Subject header, since this is already described in the Request header.

Here's an example that's a little closer to reality: The browser first establishes a direct connection to port 80 of the computer being contacted (www.bluebikes.com in this case). In responding to the original connection, both computers identify each other, so there's no need for To or From headers. Then the requesting browser sends the following:

```
GET /aboutus.html HTTP/1.0
If-modified-since: Fri, 6 Dec 1996 18:42:29 GMT
Referer: http://win22.xyz.com/coollinks.html
User-Agent:  Mozilla/3.2 (Windows 95)
```

Collectively, these lines are called the HTTP request. The first line is called the *request line.* Each line after the first line is generically called an HTTP *request header.* Notice that in this example, we've added an If-modified-since header. If a browser discovers that the document being requested is in its disk cache, it may add the If-modified-since header to the request to effectively convert it into a *conditional GET* request—where the server will only actually return the document if it has been modified since the date and time specified. If the document has not been modified, the server will return a code 304 (Not changed) status and the browser will use the copy in its local cache.

The If-modified-since header adds desirable functionality for both client and server. For the reader using the client browser, pages will load much more quickly when they're loaded from the local disk cache than when the page or graphics have to be pulled through a telephone line. For the server, fewer bytes transferred may mean many fewer dollars in bandwidth charges at the end of the month.

For both requests and responses, there isn't a fixed number of headers that must be included. The above request would work fine (from the reader's point of view) if it consisted only of the first line. Only the request line is required. The additional lines (the header lines) simply add some functionality for either the client, the server, or both.The Referrer and User-Agent headers simply permit the web server at the receiving end of the request to log that information in its log files of the same names (or in its transfer log, in the case of a combined log format).

Now let's look at a response that the web server might send back to the browser. Like the request, the response will consist of a single line, called the *status line*, plus a variable number of (optional) additional lines called *response headers* and a variable number of lines called *entity headers*. Here's how one might look:

```
HTTP/1.0 200 OK
Server: CERN/3.0 libwww/2.17
Content-type: text/html
Content-length: 2427
<Entity body>
```

This HTTP response consists of the status line (the protocol version and the resulting status of the request) followed by one response header (the Server header), two entity headers (Content-type and Content-length), and a byte stream (represented by "<Entity body>"), which is the contents of the document that was requested. The Server header relates to the web site as a whole and has nothing to do with the object that is about to be transferred to the browser. The Content-type and Content-length headers relate directly to the entity that will follow the headers—the text of the HTML document itself. The headers tell the reader's browser that the document is an HTML text file and that it contains 2,427 bytes.

Generally, response headers relate to any matter that isn't directly related to the requested entity. The three primary response headers are described in Table 7-1. Entity headers refer to, or describe, the entity that was requested and is about to be transferred. There are more entity headers than response headers. The most common entity headers are listed in Table 7-2 (For a complete list of HTTP headers, see Appendix C.). Although there are a handful of entity headers, it's rare that a server will send more than the content type and content length. In fact, many browsers won't even recognize some of the other entity headers at all.

Notice in the sample HTTP response above that there is no clear delineation in the HTTP response between response headers and entity headers. This is how it works in practice. It doesn't matter whether a server generates a header internally or, for example, a CGI (common interface interface) program generates the header. If the receiving browser is programmed to recognize the header and do something with it, it will.

Header	Purpose
Location	When a browser encounters this header, it is redirected to another URL.
Server	Specifies the name of the server software.
WWW-authenticate	Indicates that authentication will be required before the reader will be transferred the requested entity.

Table 7-1. *Response Headers*

Header	Purpose
Allow	Specifies the allowed methods (i.e.g GET, HEAD, or POST).
Content-encoding	If the entity has been encoded (as in being compressed or encrypted), this header identifies the method of encoding so that the receiving entity can determine how to decode it to its underlying content type.
Content-length	The size (number of bytes) contained by the object.
Content-type	A description of the entity itself (i.e.g HTML, GIF image, JPG image, etc.).
Expires	Attaches a time stamp to the entity after which a user agent should consider the entity stale.
Last-modified	Indicates the date and time that the entity was last modified.

Table 7-2. *Entity Headers*

For example, let's say a CGI program produces this output:

```
Content-type: text/html
Location: http://www.bluebikes.com/index2.html
```

The browser will receive the headers and be redirected to the URL referenced by the Location header. The browser won't display these lines at all, because it recognizes them as header lines. It just quietly makes a new request for the document it was referred to.

Notice that by the strict definitions we gave above for response headers and entity headers, there's one of each here. The Content-type header falls into the entity category, while the Location header is a response header. This is a good example of why the lines between response and entity headers are blurry and why all such headers traveling from a server to a client have become generically known as response headers.

Using Cookies to Track Visitors

You may occasionally hear the term "stateless" used to describe the kind of connection that takes place between a web server and a web browser. The connection is called stateless because at the core of the HTTP protocol there is no facility for continuity from one request to the next. In a sense, this is techno-speak for what we've been saying all along—that web servers have no rock-solid (or consistently accurate) way to tie a collection of hits together into the greater bucket we would like to call the visit.

But the technology is wholesome enough that clever people have devised ways to work around this feature (or perhaps shortcoming) of the technology. There are actually two methods now that have been developed enough to be in widespread use. The first and simpler of the two methods is with something called cookies. The second method involves serving all of the content on a web site dynamically through a CGI program (I'll wait until Chapter 17 to describe this second methodology).

In the last section, we mentioned that many web browsers aren't aware of all of the HTTP response headers. Headers that they don't understand they simply ignore—just as when they're rendering HTML code they ignore tags that they can't interpret. This is a good way to design software and protocols, because it leaves room for future development and new features.

Cookies are one such feature, pioneered by Netscape Communications. There's no good explanation for the name, other than it's a term that programmers have used for years to describe magic little files or magic numbers embedded in data files that allow programs to execute or allow programs to identify their own data files.

In the case of HTTP cookies, a cookie is a magic header that a server (or CGI program) sends to a client browser. Browsers aware of cookies store them on disk, and whenever the user accesses a site or directory hierarchy to which they relate, the browser sends a similar magic header back to the server.

For example, in our sample response from the server (bluebikes.com), the response header may have looked like this:

```
HTTP/1.0 200 OK
Server: CERN/3.0 libwww/2.17
Set-cookie: USERID=3456; path=/
Content-type: text/html
Content-length: 2427
<Entity body>
```

Notice the new header, "Set-cookie." This tells cookie-aware browsers to save the contents of the cookie (USERID=3456; path=/) and associate it with the domain that sent it. In this case, the server that sent the cookie is www.bluebikes.com, so the domain with which the browser associates the cookie is .bluebikes.com.

How the different browsers handle saving cookies is up to the individual browser developers. Most of them just store all their cookies in a single file called cookies.txt.

A cookie header can consist of just about any text that you could want to send back and forth between server and client. The only limitations are that a cookie be less than 4 kilobytes (you can store a whole lot of data in 4K). Other limits and minimal features that Netscape suggests are the following:

- Clients should be able to store up to 300 cookies at a time.
- A single server should not attempt to set more than 20 cookies.
- Clients should be able to handle up to 20 cookies per server.

The above response header from the bluebikes.com server will cause the browser to store the cookie that the server sent. The next request that the browser makes of the bluebikes server will include a request header that contains the content of the cookie. The original request was for the page /aboutus.html. Let's say that we follow a link on this page to the bluebikes top-level page ("/index.html" or just "/"). Here's how the request might look:

```
GET / HTTP/1.0
Cookie: USERID=3456; path=/
If-modified-since: Fri, 6 Dec 1996 18:45:22 GMT
Referer: http://www.bluebikes.com/aboutus.html
User-Agent:  Mozilla/3.2 (Windows 95)
```

Notice the new cookie request header. The server initially sent a "Set-cookie" header; once the cookie is set, the browser sends back "Cookie" headers. Thereafter, the server doesn't need to send any more Set-cookie headers to the client, but it may—to set additional cookies for special purposes, for certain portions of a web site, or to change the value of an existing cookie. Figure 7-1 illustrates this sequence of events.

Let's stop for a moment and think about the importance of this. If you use cookies cleverly by giving a unique cookie to each reader that visits your site, your HTTP transactions will gain the continuity or state information that is inherently absent in the HTTP protocol itself. Problem solved. Well, almost.

The other half of the equation is getting cookie information into your server log files and choosing log analysis software that makes use of cookies.

CLIENT (Browser)	**SERVER (Web Server)**
Request Headers	**Response Headers**
GET /aboutus.html HTTP/1.0 If-modified-since: Fri, 6 Dec 1996 18:42:29 GMT Referer: http://win22.xyz.com/coollinks.html User-Agent: Mozilla/3.2 (Windows 95)	
	HTTP/1.0 200 OK Server: CERN/3.0 libwww/2.17 **Set-cookie: USERID=3456; path=/** Content-type: text/html Content-length: 2427 <Entity body>
GET / HTTP/1.0 **Cookie: USERID=3456; path=/** If-modified-since: Fri, 6 Dec 1996 18:45:22 GMT Referer: http://www.bluebikes.com/aboutus.html User-Agent: Mozilla/3.2 (Windows 95)	
	HTTP/1.0 200 OK Server: CERN/3.0 libwww/2.17 Content-type: text/html Content-length: 1876 <Entity body>
GET /products/index.html HTTP/1.0 **Cookie: USERID=3456; path=/** Referer: http://www.bluebikes.com/ User-Agent: Mozilla/3.2 (Windows 95)	
	HTTP/1.0 200 OK Server: CERN/3.0 libwww/2.17 Content-type: text/html Content-length: 2245 <Entity body>

Figure 7-1. *A sequence of HTTP requests and responses showing how cookies are set, transferred, and changed*

Configuring the Server to Send Cookies

If you've already chosen Netscape servers for your web server software, you're in good shape. Netscape's support for cookies is the strongest in the industry. This isn't surprising, since they pioneered the use of cookies. If you use a Netscape server, you can choose a configuration option to log cookie information in the transfer log.

If you use other servers, configuring them to send, respond to, and log cookies isn't nearly as easy. The Apache server for Unix systems includes a source code module for implementing cookies. You can also modify the log reporting module to include the cookie information on each line of the transfer log. If you go to the trouble of modifying the logging module to report cookies, you might as well also include the referrer and user agent information in the log at the same time. The way this works is that you modify the LogFormat string in the source code to look like this:

```
LogFormat "%h %l %u %t \"%r\" %s %b \"%{Referer}i\" \"%{User-agent}i\"\"%{Cookie}i\""
```

Intersé, one of the log analysis software leaders, has taken an active role in helping to move this forward. On its web pages, Intersé offers (more or less) drop-in modules for several servers, including the Apache server, Microsoft's Internet Information Server, and the Netscape servers. For the Unix servers, these modules are C source code that you must integrate into the C source for the server and compile. For the NT servers, the modules are new DLL files and instructions on updating your system registry to use them.

Setting Cookies with CGI

Since a cookie is set simply by including a response header, it doesn't necessarily have to be the server software itself that sets cookies for your site. With a little imagination, you can probably come up with a myriad of ways to set cookies without a single modification to your web server.

For example, an image on your top-level page (or on every page, for that matter) can reference a CGI script instead of referencing an image file directly. The CGI script can simply check to see if the request included a cookie header. If it didn't, it can generate a Set-cookie response header before sending the content of the graphic file.

Setting cookies this way is fine for some applications, but it's not particularly good for tracking visits. Let's say you want to use cookies to track state information for a shopping basket application. A reader enters your "virtual store," which trips a CGI program that sets a cookie for the visit. If the reader marks a product for purchase and clicks a form button labeled something like "Add to shopping cart," the CGI program that gets the form data adds the product code (and quantity if applicable) to the existing cookie that it sends back to the browser after each request (replacing the previous version of the cookie with a new one that includes the additional product code and quantity). When the reader finally clicks on a button indicating that the shopping spree is complete, this final CGI program gets the full cookie, which includes the IDs of every product the reader wants to buy.

This is a special application for which CGI-set cookies are well suited. Setting cookies with a CGI script, however, doesn't get your web server to log cookies in your transfer log. Sure, you can make CGI scripts that set and read cookies to do their own logging in as elegant a way as you like. But that still doesn't get cookie data into your transfer log where you need it for tracking visitors.

A small percentage of the log analysis software and services support cookies in the transfer log today. That number is growing quickly as developers realize that cookies may be the best way to extend the web and the HTTP protocol to track state information about connections. It's altogether possible that by the time you read this, most—or even all—of them will support cookies.

Security Issues with Cookies

The reputation of cookies was tainted early on in a firestorm of magazine articles and discussion groups on the Internet relating to security concerns. It's unfortunate that cookies were associated with this and that as a result cookies are often thought of in a negative way.

The trouble is not really with cookies at all but rather with executable code extensions like Java, JavaScript, and ActiveX. These technologies are HTTP extensions that allow web browsers to download and execute programs on your computer—with or without your knowledge. Java and JavaScript are the most common of these extensions, and you usually have some clue when they're running. They're used for displaying marquees, rotating ads, showing 3-D pictures, or a myriad of other things (in fact, practically anything that a regular computer program could do on your computer). Now, *this* is inherently dangerous. At the extreme, some nefarious person could write programs that delete files, look for private information and send it back to the remote computer, or propagate a computer virus.

At some point, some innovative—and unscrupulous—person had the idea of using a Java applet or JavaScript to look up readers' e-mail addresses and embed them into their own server's cookie. Thereafter, when readers accessed another page from the server, they unwittingly sent their e-mail address to the server in the text of their cookie.

It was by this association that the reputation of cookies became tainted. In reality, such a Java applet or JavaScript script could have used many methods to get these e-mail addresses back to the offending site, including e-mail or even opening a direct TCP connection to send the information. The cookie just happened to be the method that this unscrupulous person chose—making the cookie an unwitting dupe.

A second objection sometimes recited about cookies is the abject obtrusion of remote web sites that set cookies on their readers' computers. Some feel that it's a violation of their own resources to be forced to store information on their computer for the benefit of remote sites.

This objection doesn't really hold any water, either. Although it's true that web browsers are supposed to be able to handle up to 300 total cookies of up to 4K each (a theoretical maximum of about 1.2MB), it's rare that a cookies.txt file will exceed a couple of kilobytes. Most cookies come nowhere near 4,000 bytes (or characters) of text—they're usually just a few dozen characters—and it's far more likely that even an active web surfer will have closer to a few dozen cookies (not 300). Moreover, most people store between 5MB and 10MB of HTML pages and graphics in their disk cache. The amount of disk space is trivial.

And while there may be some truth that cookies benefit us (the people running web servers) more than they do the readers, we should try to remind people that cookies will allow us to tailor more desirable—and more effective—web sites. The benefit to readers isn't short-term; in the long run, cookies will benefit both site developers and readers.

A third objection to cookies is the information they store. Unsophisticated web users can have a paranoia about giving away information about themselves, and justifiably so. What they often don't understand is that a cookie doesn't allow the server to know any more about readers than it already knows (the type of browser they're using, the domain they come from, the referrer page, and so on).

Some people do use cookies for more. For example, in the previous section we mentioned how some "shopping basket" applications use cookies to pass the product ID numbers of products that a reader buys. This way, when the reader wants to "check out," the server will know what products the reader wants to buy. This information is pretty benign. The cookie will be storing nothing sensitive—just some product IDs. More than likely, the final CGI program to process the purchase will reset the cookie to clear out this information.

Perhaps an example of something shady that you could do with a cookie is to take some information that a reader enters into a form on a web site (like a credit card number, address, or e-mail address) and store it in the cookie. Even that wouldn't be anything that the reader didn't voluntarily give to the web site freely in the first place, and it can only be used when communicating with that particular site. So it would just be the same information flowing back and forth between the same client and server. Nobody would do this anyway—it makes no sense. If people want to get and keep information on a reader, as with a user registration system, they will prefer to store it on their own computer and not in a cookie.

There's not really any motivation to turn cookies off—they're virtually completely benign. All they really do is allow the web server to associate hits from the user's browser with previous hits (and visits) from that browser. With Java and JavaScript, on the other hand, there is the potential for violation. Java applets and JavaScript scripts can do some nasties with or without cookies, which is a potentially large motivation for people to turn off Java and JavaScript. (To disable JavaScript, just uncheck the "Enable JavaScript" box in a browser's options screen.)

Despite the facts, some people will still ask how to turn off cookies. The answer is that there really is no way to disable the acceptance of a cookie. The only thing you could do is delete the cookies.txt file before running the browser, but, again, there's really no motivation to do so. Some magazine article authors have suggested that if you don't want to accept cookies, get a browser that doesn't support them; they have also suggested that browser vendors should let users turn cookies off.

The two largest and most important browser developers are 100 percent behind supporting cookies, and they don't provide any way for users to turn them off. I expect that other developers will follow suit and there will be fewer and fewer browsers that don't support cookies. I agree that there should be no way to turn them off—there's no compelling reason for it.

Complete paranoids can always get around dropping cookies by using a service such as the Anonymizer. I doubt this will ever happen in a big way. Imagine getting mad at the phone company because a phone call could be traced. That doesn't mean that every call you make gets traced. And with a phone number, you have the potential for much more information about someone than you have with a cookie. In summary, cookies pose no threat to anyone. As is the case with many things, unscrupulous people will use the tools at their disposal to commit dastardly acts. But there are far more dangerous tools than cookies.

In Chapter 8, we will look at ways to use your site statistics to make your web site more effective. Then we will jump into Part 3 for head-to-head comparisons of commercial and shareware log analysis tools and services, and we'll discuss some of the issues you'll encounter if you are contemplating developing your own log analysis software.

Chapter Eight

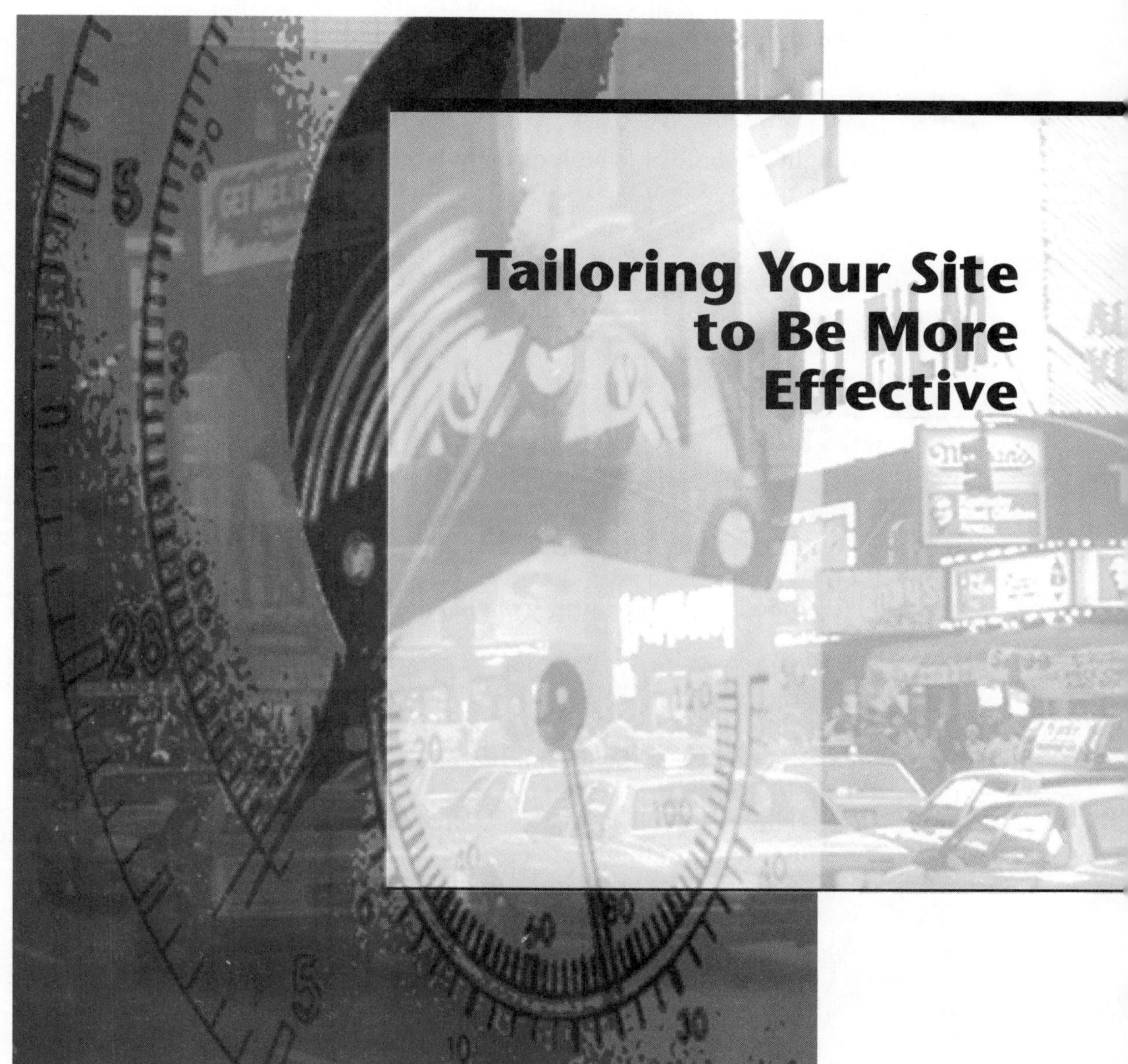

Tailoring Your Site to Be More Effective

In the last chapter, you learned about one concrete way to track visitors through a web site—by using server-set cookies. Also, for the first time in the book, we showed you some of the statistics that you can track for your web site whether or not you use cookies. In this chapter, I will give you some tips and pointers for crafting a more effective web site.

The process of reflecting on traffic statistics and making content and structural changes to your web site based on those statistics is a little bit science, a little bit voodoo, and whole lot of guesswork. In short, it's highly subjective, nebulous, and even more situational.

Nevertheless, I will attempt to go beyond the often-repeated obvious to shed some light on some of the things that your server statistics can reveal about how people use your web site and actions you can take to modify their behavior.

Before we get into ideas on adapting sites to be more effective, we need to look at how to analyze summary statistics. We'll start off by focusing on that, then we'll look at identifying your site objective, then on to trimming and hemming to fit.

Analyzing Summary Statistics

In Chapter 6, we followed the tracks of a single visitor through a web site. We looked at the path that a visitor we called Bob followed through the site, and we came to a fairly solid understanding of how he found the site, what he was looking for, and what he looked at instead of what he really started out to find. We also mentioned the impracticality of analyzing individual visits at this level of detail.

Unfortunately, as of yet, most of the commercial (and freeware) log analysis packages and services don't go to the level of detail that we went to in dissecting that visit. For example, we factored into our analysis the time that it took for each page to finish loading the text and all of the graphics associated with each page. This hasn't been a priority for developers of log analysis tools, partly because the ability to do it is dependent on the particular server the site uses and its run-time configuration.

We have devoted a good deal of ink in this book to transitioning the focus of log analysis from the hit to the visit. Now, with cookies, we've arrived at the visit, and we have to bear just a little bad news about analyzing it in summary. Even if you have configured your server (say, by using cookies) and log analysis software to track visits accurately, you're going to lose some information when you summarize visit data. To use our earlier visit in Chapter 6 as an example, recall that Bob followed this path through the web site:

Time	**Duration**	**Page Viewed**
15:16:19	33 sec	/notfound.html – error document
15:16:52	1 min 14 sec	/index.html – the top-level page
15:18:06	3 min 48 sec	/tips/qa2.html – the tips/Q&A page
15:21:54	56 sec	/tips/index.html – the main tips page
15:22:50	1 min 30 sec	/news/index.html – the news page
15:24:20	42 sec	/proshop.html – the pro shop
15:25:02	18 sec	/proshop/balls.html – the balls page
15:25:20	42 sec	/proshop/titleist/titlball.html – Titleist balls
15:26:02	44 sec	/proshop/titleist/bridgestone.html – Bridgestone balls
15:26:46	unknown	/notfound.html – error document

Now let's say that another visitor—we'll call him Jim—followed this path through the web site:

Time	**Duration**	**Page Viewed**
11:46:24	1 min 33 sec	/index.html – the top-level page
11:47:57	1 min 25 sec	/news/index.html – the news page
11:49:22	1 min 3 sec	/leaderboard.html – the leader board
11:50:25	2 min 25 sec	/proshop.html – the pro shop
11:52:50	20 sec	/proshop/irons.html – the irons page
11:53:10	1 min 10 sec	/proshop/cobra/cbraims.html – Cobra irons
11:54:20	2 min 16 sec	/proshop/cobra/cbratour.html – Cobra tour irons
11:56:36	2 min 19 sec	/class/ – Classifieds main page
11:58:55	3 min 22 sec	/class/browse.html – Classifieds browse page

Time	Duration	Page Viewed
12:02:17	1 min 27 sec	/class/clssirns.html – Classifieds irons page
12:03:44	1 min 42 sec	/class/clssirns/msgs25791.html – a classified ad
12:05:26	unknown	/class/clssirns/msgs14597.html – a classified ad

Jim entered the site in a somewhat more conventional way—through the front door. He checked the news, the leader board, then he dropped into the pro shop to check out the irons. Finally, Jim checked the classified ad section for irons as well.

Let's think about how we would summarize these two visits. Ostensibly, there isn't much difference between summarizing two visits and summarizing 10,000 visits. The exercise is the goal here—to think about *how* you would summarize detail visit data.

Here's a thought: we could come up with a statistic that identifies what pages people tend to look at first and last in their visits. No doubt, most people will begin their visit with the top-level page of the web site, but deviations from that could be both interesting and informative. Of course, it's not all that interesting with only two visits to look at. Here, 50 percent of our visitors entered through the top-level page, while the remainder began their visit on the 404 (Not found) error page. Some analysis software—like Intersé's Market Focus—give you the top ten entry points to your site listed in descending order (Figure 8-1). This table also shows you the number and relative percentage of visits that began on each of these pages.

If you know your site well, your top entry pages will probably make sense to you. That is, you won't be surprised by the pages through which people enter your site. They're likely to be some of your most popular pages. Take advantage of this knowledge by trying to draw people from those pages to where you want them to go. If you don't already advertise on these pages, experiment with putting a banner ad on them to draw them to your sweet spot. At the very least, be sure to have buttons and links here that link to the places on your site that you would like your visitors to visit.

Another statistic that comes to mind is the average time spent on each page. This one raises some interesting questions. How can you do this computation when only a fraction of all visits have actually viewed a

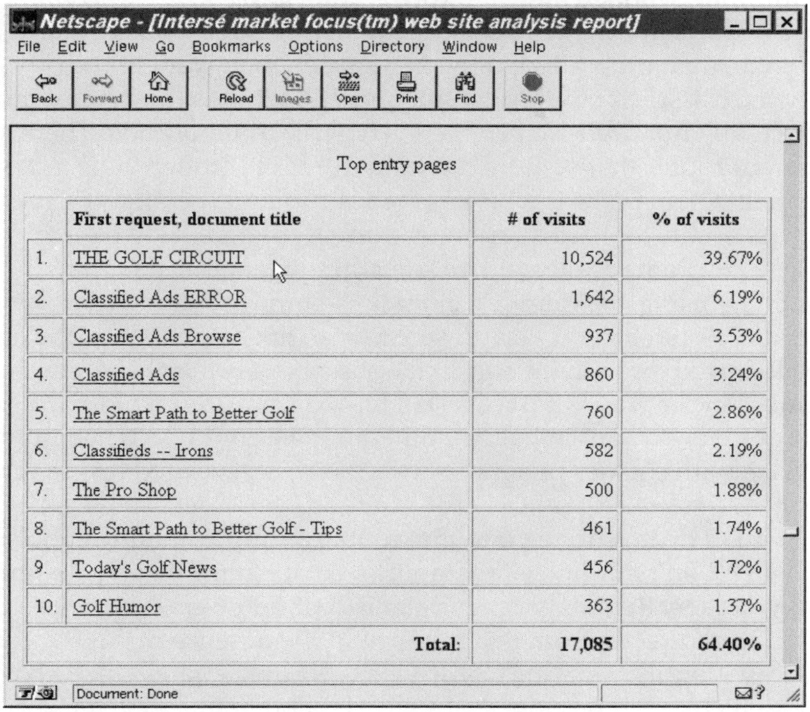

Netscape - [Intersé market focus(tm) web site analysis report]

Top entry pages

	First request, document title	# of visits	% of visits
1.	THE GOLF CIRCUIT	10,524	39.67%
2.	Classified Ads ERROR	1,642	6.19%
3.	Classified Ads Browse	937	3.53%
4.	Classified Ads	860	3.24%
5.	The Smart Path to Better Golf	760	2.86%
6.	Classifieds -- Irons	582	2.19%
7.	The Pro Shop	500	1.88%
8.	The Smart Path to Better Golf - Tips	461	1.74%
9.	Today's Golf News	456	1.72%
10.	Golf Humor	363	1.37%
	Total:	**17,085**	**64.40%**

Document: Done

Figure 8-1. *Intersé's top ten entry point table*

particular page? In this example, the two visits intersected on only three pages. The common pages are

```
/index.html - the top-level page
/news/index.html - the news page
/proshop.html - the pro shop
```

An average page view really only takes on any special meaning if a significant portion of visitors have actually viewed the pages for which you do the computation. This isn't likely to be a problem—you will surely have more than a few visits in your log files when you run your statistics.

The top entry pages and time spent per page are two statistics that come easily to mind that you can track if you are monitoring visits actively. Let's assume for a minute that we have used cookies or some other method to track visitors positively. What do we really have to gain from it when we summarize the data?

The paths that people will take through your web site are too numerous, so it would be difficult to track statistics like "How many visitors started on the home page, then proceeded to the news page, then went to the pro shop, and finally the classified ads?" Most fully functional sites that attract much traffic have many hundreds of pages—even perhaps thousands. There are so many different routes that visitors can (and do) take through such a site that reporting their actual paths is a bit impractical. To date, no analysis packages offer statistics on the top paths followed through the site. However, it's not impossible to do. Perhaps this will be something we will see in the future.

Although by using cookies we can track visitors' absolute paths through a web site, there's not a good way to summarize the detail data of the actual paths they followed through the site. In the absence of being able to summarize this data in a concise way, we have to fall back on statistics that report simply the number of views for each page—something we could do whether we were tracking visits with cookies or simply analyzing unassociated hits in the transfer log.

So what good does it do to go to the trouble of implementing cookies and tracking visits? Don't minimize the importance of having a concrete number of visits to your site. Remember that a significant number of statistics you will want are derived directly from this number. In the absence of having cookie data to solidly identify visits and return visits, your statistics will be based on a guess, which might be close, but you can bet that it won't be right.

Identify Your Objective

If you want to improve the effectiveness of your web site, one of the first things to think about is your objective with the site in the first place. There are basically three broad categories:

- *To establish an organizational presence on the web:* This may include informing the public or potential customers about your company, making them aware of your products or services, and servicing the existing customer base through web-based customer service and technical support. Although any company may sell its products or services through its web site, if the primary objective of the site is not selling, then it will probably fall into this category.
- *To sell products through the web site:* This web site may, in addition, offer customer services and technical support. The emphasis is important: if

the primary objective is to sell products, the objectives for tailoring the site will be different.

- *To sell advertising on the site:* The prima facie objective of such a site is to provide compelling and continually new content to attract as many new and return readers as possible. The purpose is to maximize ad exposure for advertisers.

Each of these three categories has its own relatively unique requirements for design, content, and tracking. Not coincidentally, the perceived importance of tracking site traffic and tailoring content increases exponentially through these three categories, from the first to the third.

Customer Service and Promotional Sites

For companies or organizations whose objectives fall in the first category, simply having a web site that is visually appealing, that works, and that draws traffic may be enough. These sites are plentiful on the web. In fact, these sites may represent the majority of all commercial sites. For example, take the RCA web site at www.nipper.com (Figure 8-2). In it, RCA provides detailed information about all of the products that it makes and sells. If you're looking for information on the latest and greatest digital video camera or if you're trying to figure out how to hook up your new satellite system to your TV and VCR, this web site has the info and hookup diagrams for you. It also provides technical information, such as wiring diagrams, frequencies, and specifications.

RCA does not, however, sell its products through its web site. Neither does it advertise for other companies. In fact, it would be quite odd to see an advertisement for another company on such a web site.

To what extent does RCA analyze its server log files? They may analyze them thoroughly, but then again, they may not at all. No doubt, they check the error log daily to flesh out problems. And if they do run statistics, the reports probably make their way up through several layers of management—if only to let them justify the expense. But everyone involved knows that the benefits of this web site aren't completely tangible—or even measurable.

Nevertheless, it's a good idea to look at your server statistics closely, because what you find may not jive with what you expect. For example, areas of your web site that you think are particularly interesting might have little or no traffic. The reason could be that your readers simply don't know about this

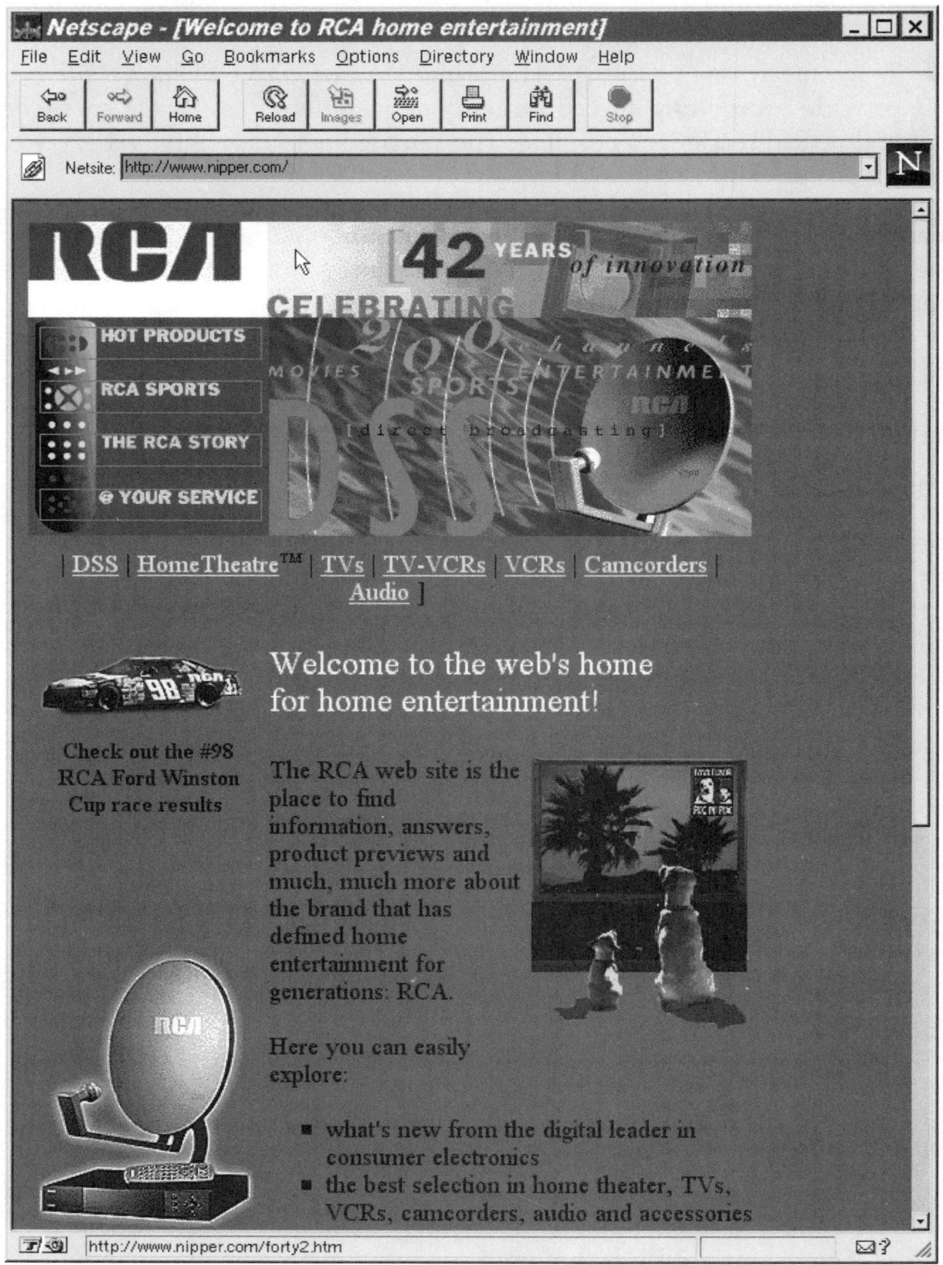

Figure 8-2. *RCA's web site promotes its products and supports customers*

area of the web site. If you think that it is important for more people to find this area, try putting more hypertext links to the area from pages that are more heavily trafficked. You could also put banner ads or Java applets on your pages to draw more attention to the resource you're trying to get noticed.

Increasingly, log analysis software and services are recognizing the need to analyze page cancellations. This is when readers stop a page transfer in midprocess, either by clicking on another link they see on the page or by clicking Stop or their back arrow. As we mentioned earlier, it's tough to tell if cancellation occurs because someone clicks Stop or clicks the back arrow. This is because the previous pages that the reader viewed are generally in the browser's memory cache—which means that if they return to a page they have viewed previously, there's no indication at all in the log files—not even a status 304 (Not changed). Of course, cancellations can also be caused by something mechanical like the reader's modem dropping its connection.

NetCount's log analysis service is one of the leaders in recognizing the importance of reporting cancellations. NetCount makes a clear distinction up front between what it calls BIRs and BITs. A BIR is a browser information request, and a BIT is a browser information transfer. These statistics are similar to what NetCount calls PIRs and PITs: a PIR is a page information request, and a PIT is a page information transfer. Sound confusing? Basically, BIRs and BITs refer to all requests and transfers, and PIRs and PITs refer to HTML pages only.

Figure 8-3 shows NetCount's BIT/PIT report for the golf site. Even though you can't see the report for every hour of the day, you can see enough of it to realize that PIR/PIT success rates generally run higher than BIR/BIT success rates. This makes sense—the text of pages tends to load up and display well before the graphics all transfer, and people tend to click on links they can see and are interested in and aren't concerned with waiting for graphics to finish their transfer.

Some of the other services and software don't report on cancellations at all—at least not yet. But the competition in this field is fierce right now. Don't be surprised if they all report cancellations to some extent by the time you read this.

If the software or service of your choice does offer a cancellation (or request success) report, study it carefully. This can give you clues as to which pages are too graphically intensive or are too large, which may explain a high cancellation rate—people just don't want to wait around for your graphics and Java applets.

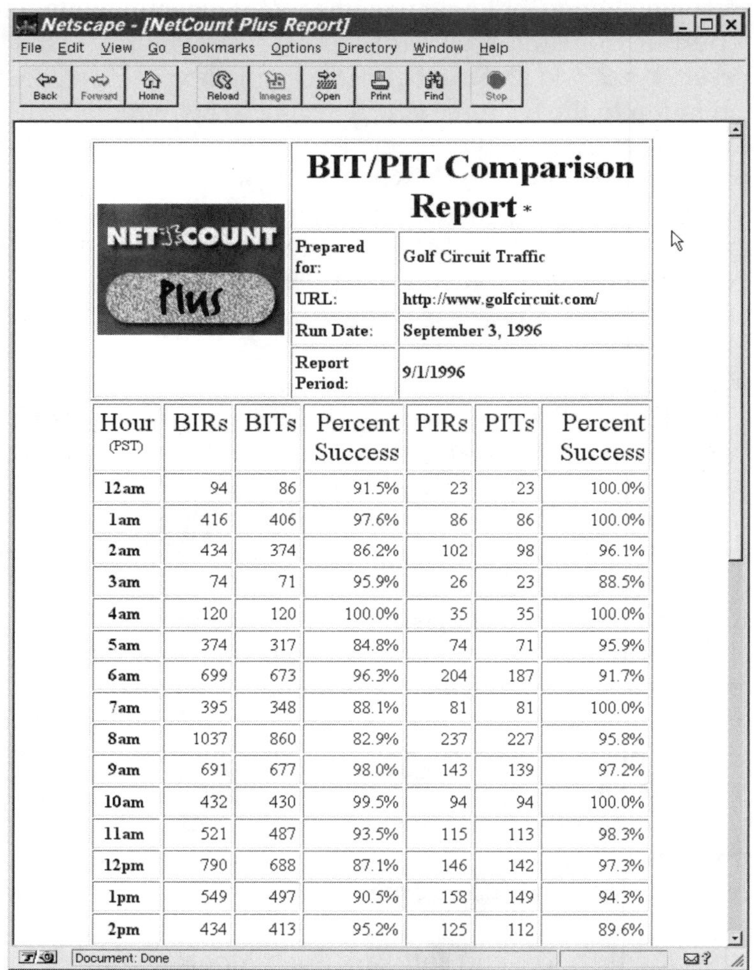

	BIT/PIT Comparison Report
Prepared for:	Golf Circuit Traffic
URL:	http://www.golfcircuit.com/
Run Date:	September 3, 1996
Report Period:	9/1/1996

Hour (PST)	BIRs	BITs	Percent Success	PIRs	PITs	Percent Success
12am	94	86	91.5%	23	23	100.0%
1am	416	406	97.6%	86	86	100.0%
2am	434	374	86.2%	102	98	96.1%
3am	74	71	95.9%	26	23	88.5%
4am	120	120	100.0%	35	35	100.0%
5am	374	317	84.8%	74	71	95.9%
6am	699	673	96.3%	204	187	91.7%
7am	395	348	88.1%	81	81	100.0%
8am	1037	860	82.9%	237	227	95.8%
9am	691	677	98.0%	143	139	97.2%
10am	432	430	99.5%	94	94	100.0%
11am	521	487	93.5%	115	113	98.3%
12pm	790	688	87.1%	146	142	97.3%
1pm	549	497	90.5%	158	149	94.3%
2pm	434	413	95.2%	125	112	89.6%

Figure 8-3. *NetCount's BIT/PIT report*

Product-Oriented Sites

A minimal overview of server stats may be enough for a company or organization simply maintaining a presence on the web. But for a company depending on making product sales or selling ad space for its livelihood, the statistics can be vitally important to know how people are reacting to certain aspects of the site—why they look at the pages they do, why they don't look at other pages, and why they make their decision to buy or not to buy.

Take, for example, Stout's Cider Mill at www.cidermill.com (Figure 8-4). (Yes, the name is no coincidence. It's my dad's site—created and maintained by your author.) Stout's Cider Mill is a bustling business in the Southwest. It sells apple and other fruit and nut products (pies, cakes, cider, jams, and so on) through its stores and on the road in three states. The objective of this web site is to sell products. Every page is geared toward getting people to buy products online or at least to get them to go to the stores.

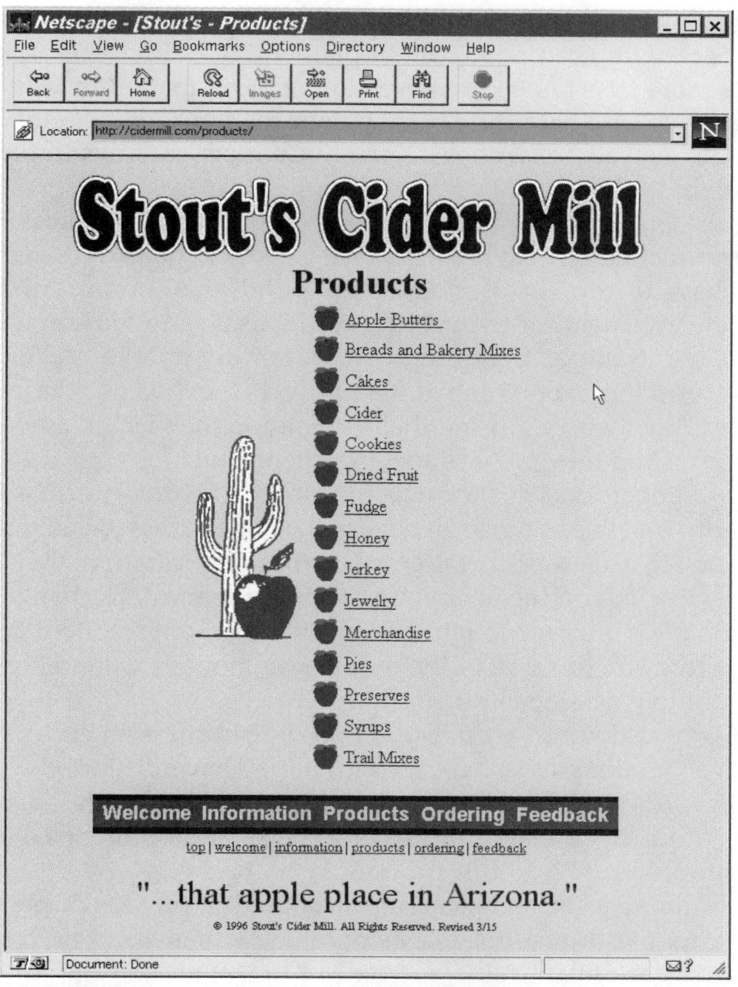

Figure 8-4. *Drawing visitors to the order page is the goal of many web sites*

In this case, the web site business augments the real world, day-to-day business of the stores. I run statistics each month with a variety of log analysis software packages to examine the traffic through the web site and to come up with ideas to draw more people into placing orders.

The most obvious thing you want to do on a product-oriented site is to draw people either to your product pages or to your order form or a page where they can initiate a shopping spree by starting a shopping basket application.

The first issue this brings up is whether to use a one-page order form or to use a shopping basket. The answer is simple: you want to use a shopping basket approach if you can pull it off. Implementing a shopping basket takes considerable more time and skill than a simple order form. Actually, this is one reason it's good to do a shopping basket: your readers will be more impressed with your skill level and expertise if you have a more complex implementation of a shopping basket than if you have a simple order form.

Another reason to use the shopping basket approach is to make buying easier for your readers. If you have separate product and order pages, your readers will have to jump back and forth to "build" their order form. This may not only be inconvenient for them, but it could also be downright not worth their trouble. For example, if they start an order form only to find out that they need more information about one of the products they're considering buying, then they may feel they risk losing the information they've already entered into the order form if they go back to look at a product.

One way to get around this to some extent with an order form is to put a link on each product line in the form back to the page with detailed information about the product. This way, if readers need more information, they can simply follow the link on the form to read about the product, then click the back arrow on their browser to return to their form. Some readers will still be hesitant to follow any links off of the order form, however, not realizing that they can return to it as they left it.

The best way to do product sales is to provide buttons on each product page, effectively making each page a form. If they haven't started a shopping basket yet, the button can be labeled something like "Start a shopping basket and add this product." If they have already started a shopping basket, the button can simply read "Add this product to my basket."

Obviously, doing a shopping cart application this way requires state information about your readers. In other words, you need either to be serving your product pages dynamically (and embedding some sort of hidden code or field in the forms to allow you to identify unique users) or using some other facility to provide state information such as cookies. Using cookies is the simpler way, but serving the content dynamically may still be preferable (we'll look at this in Chapter 17).

Even on a product-oriented site, the tips we went over in the previous section still apply. Look at the relative traffic on each of your pages. Try to draw users to the product pages or to a page where they can start a shopping spree. Use banner ads, extra links, Java applets, and JavaScript scripts to draw attention to the bargains and excellent service you intend to provide your customers.

Advertising Sites

The third objective of a web site deals with selling advertising. The golf site we've referred to in this book is such a site. The Golf Circuit (www.golfcircuit.com) is for anyone interested in the sport of golf (Figure 8-5).

Being a special-interest (or affinity) site, The Golf Circuit attracts a lot of visitors. Many new visitors find the site every day, but most of the traffic is return visitors dropping in for the latest news and developments in tournaments and on the golf circuit.

The major search engines, such as Yahoo and Infoseek, are good examples of general-interest sites that sell advertising. Being of general interest, these sites tend to have traffic several orders of magnitude greater than special-interest sites. There are, however, fewer of them. No doubt, we'd all love to come up with the next new general-interest idea that will attract the masses, but novel new applications are getting harder and harder to come by.

Whether it's a relatively small special-interest site or a large general-interest site, the goal of this type of web site is to display as many pages (and ads) as possible. The search engines have a built-in advantage here. Since they are primarily serving HTML generated on the fly, it's easy for them to deliver highly focused ads where users are drilling down into specific areas of the sites. This format also makes it especially easy to deliver ads selectively from a playlist to maximize the number of ad exposures.

Smaller special-interest sites have their advantages too, though. Focusing ad relevance is natural on a special-interest site. If you have a site devoted to sewing or playing chess, Gatorade probably shouldn't be on your list of prospective advertisers. It's in everybody's best interest to make sure that the audience is right before even considering running advertisements.

If your objective is advertising, your goal is to attract as many users as possible. Since, generally, advertisers pay by the ad view (which is called an impression), you already know that displaying an ad as many times as possible (or to fulfill an ad contract) is how to meet your objective.

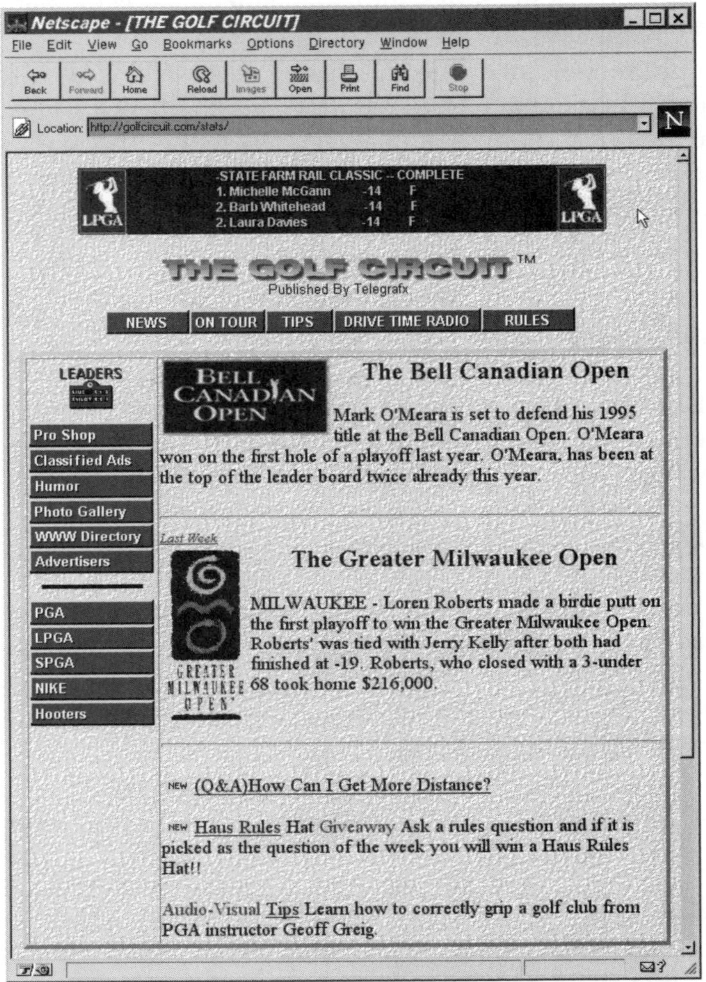

Figure 8-5. *General (or special) interest sites most often hope to sell ad space*

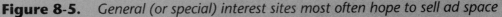

When web-based advertising was first getting off of the ground, some advertisers got excited about the prospect of paying by the click-through rather than by the view. This is understandable. The genesis of their thought process led them to realize that they could do something with web-based advertising that they have never been able to do with any other medium: track referrals to their source and actually track the effectiveness of specific ads.

However, further thought shows the click-through rate to be an undesirable basis for compensation—at least by itself. In an advertising agreement, it is generally the responsibility of the site manager (or webmaster) to attract the impressions and provide adequate placement. A click-through rate says more about the ad audience and the quality of the ad itself. If the ad audience is good (e.g., running ads for golf clubs or golf bags on a golf-related site), the responsibility for the click-through rate falls squarely on the quality of the ad banner itself.

How can an advertiser expect its medium (the company or person maintaining a web site) to pay for something over which they have no control? The answer is that they can't. Some ad banners are simply more effective than others. You've probably seen ad banners that have given you absolutely no desire to click on them, and you've no doubt clicked through many others.

With advertising-focused sites, there's some level of disconnection between the content of the site and the site statistics. With this type of site, there's no doubt that the single most important statistic is how many views there were on each page. The disconnection comes into play because drawing readers to any particular page is nearly an art form. In reality, you probably have some idea of what your special-interest readers are going to be interested in, but every page you do will be a crapshoot; only time will tell if it will become popular and will be read by many people.

Pages that seem to always be at the top of people's list of interests are typically news pages, current event pages, "what's hot" pages, and classified ad pages. Often, reader-written classified ads will be some of your most popular pages. Don't miss an opportunity to advertise on these pages.

The other difficulty in advertising on a site is in knowing when to declare that enough is enough. If you have only two or three advertisers—and 300 pages—you don't want to be running the same ad banners on every other page that your readers view. They'll quickly grow tired of seeing the same banners over and over again. Then again, marketing types would argue that maybe their click-through rate will really begin to take off after people have been exposed to an ad many times. Some actual studies have indicated that click-through rates decline sharply after two or three impressions, so after a few views, you might as well try something else.

We'll go into much more detail about advertising on the web in Part 4—there's much more to it than we've covered here. But hopefully, in this chapter, we've given you some food for thought on how to go about using your server stat reports to craft a more effective site.

PART THREE

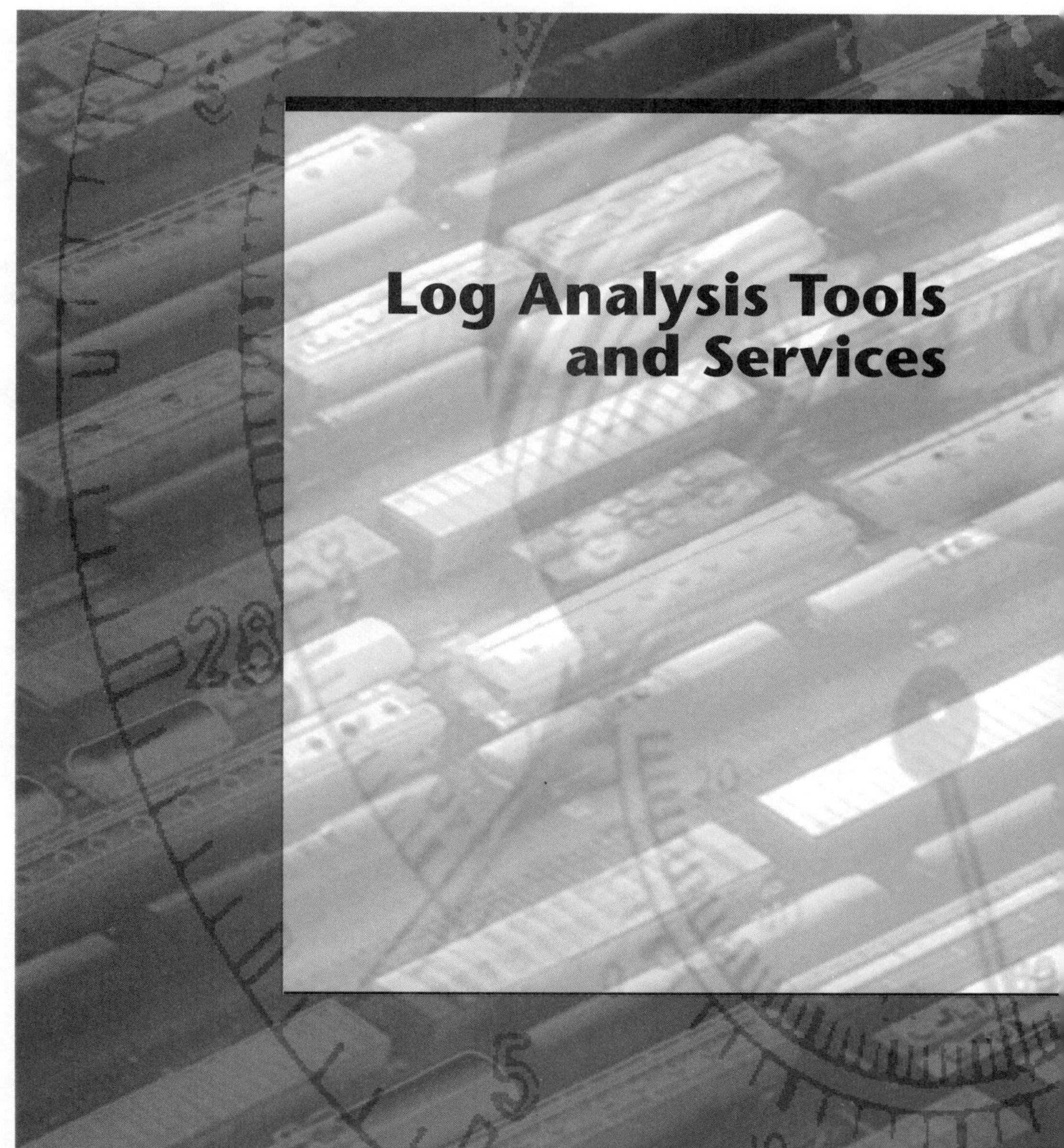

Log Analysis Tools and Services

Chapter Nine

Commercial Log Analysis Software

In this chapter, we will finally examine several of the most popular and important commercial software packages for web server log analysis. We will take an in-depth look at the three products that have Windows NT or Windows 95 versions—or, at least, Windows client versions.

Evaluation versions of each of these products are on the CD-ROM included with this book. If you want to be sure that you get the latest and greatest version, you can download them directly from the web sites of each developer. (We also include hypertext links to these companies on the CD.)

WebTrends

WebTrends is a product of e.g. Software, Inc. of Portland, Oregon. Of the commercial log analysis packages, WebTrends is probably the most commercial—by this I mean that it enjoys the widest distribution and is one of the most heavily advertised log analysis products.

One of WebTrends' biggest plusses is its simplicity. You control the entire program from one small program window with a list of your log file setups and a toolbar for all of the features you will want to use (see Figure 9-1).

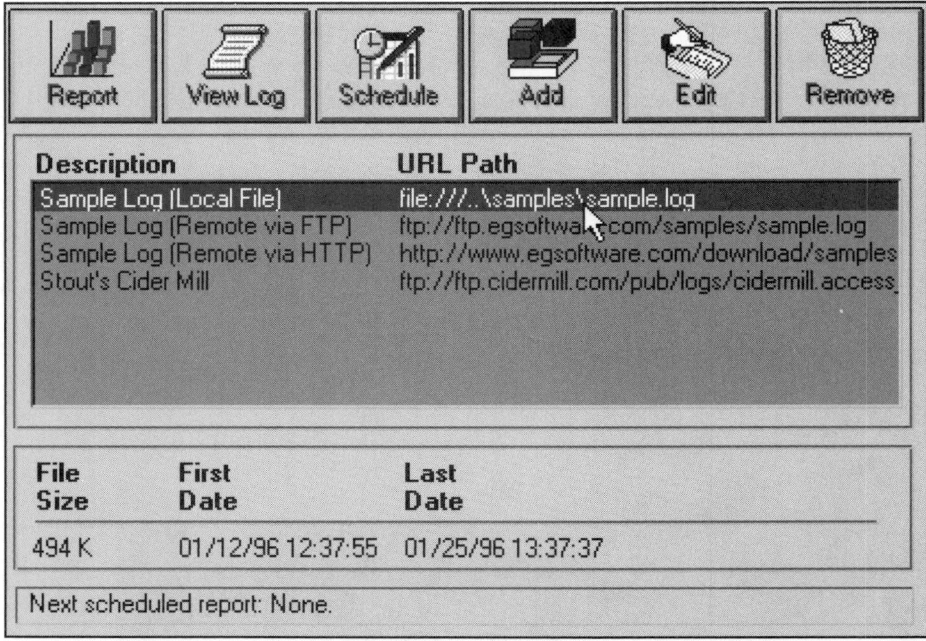

Figure 9-1. *WebTrends is a popular commercial log analysis package available at most software stores*

Developer:	e.g. Software
Platforms:	Windows 3.1x, Windows 95, Windows NT
Evaluation copy on CD?	Yes
Evaluation copies online?	Yes
Web site:	http://www.webtrends.com
Online purchase?	Yes
Retail price:	$299

Data Import

The first thing you have to do to set up WebTrends is define a new Log File Entry. This is where you tell WebTrends how to retrieve (or where to find) your server access logs. Figure 9-2 shows the Add/Edit Log File Entry dialog box. The entry fields are simple—and there are only six.

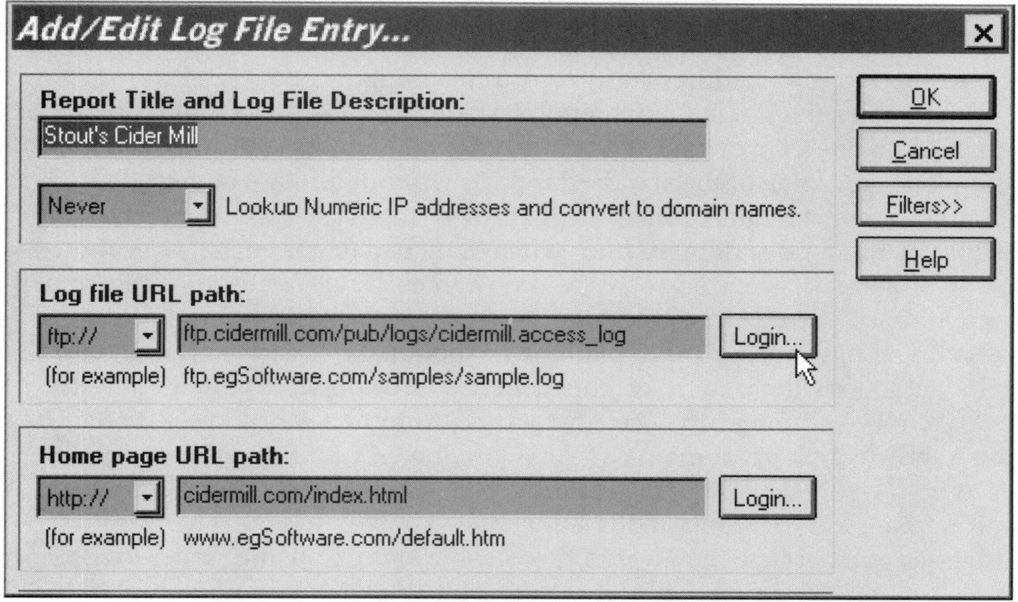

Figure 9-2. *The WebTrends Add/Edit Log File Entry dialog box*

Report Title

The first field is for the title, or name, that you want to give your particular entry. If you are using WebTrends to produce statistics reports for multiple sites, you will want to set up an entry for each site. But there's no need to do that all at once. You can set them up one at a time as you need to in order to run each report.

Although you can define as many entries as you want (even for the same site), if you plan on automating your imports and report runs, you will probably want to settle on only a single entry for each site that you automate. The reasoning is that if you plan to automate the process, you probably intend to automate on both sides. On the computer with the web server that produces the logs, you will have a routing maintenance script that rotates and archives logs—and places the current complete log in a consistent place so that WebTrends can find it each time it runs.

Domain Names

The second field tells WebTrends whether to look up the domain names of log entries that have only an IP address in the host field. The value you put in this field depends on a couple of considerations. In Chapter 2, we mentioned that it is a good idea to turn off domain name lookups in a web server if it is at all heavily loaded. If you have done this, you will have no domain names; rather, you will have only IP addresses in your transfer log.

When you run your statistics reports, it is much nicer to look at reports that refer to domains than ones that work only with IP addresses. Moreover, it isn't possible to run some of the standard reports without domain names. For example, a report on the number of visits by domain type breaks down your visitors by the type of domain from which they hail (commercial, educational, and so on); unless WebTrends has domain names, it can't determine the type of domain represented by an IP address.

So if you have turned off domain name lookups in your web server, you will want to look them up now. This field has three possible values: Never, Always, and Automatically. The Always and Automatically choices direct WebTrends to look up domain names. If you haven't configured your web server to turn off domain name lookups, odds are that most of the entries in your transfer log will contain domain names. Those entries that have only IP addresses will because they weren't resolvable at the time the server logged the web access (and probably still aren't resolvable). If you your server is configured to look up domain names when it can, you will want to make sure that Never is your choice in this field, because doing domain lookups will increase your report generation time exponentially.

Log File Path

The next two fields relate to the path or URL for the log file you want to import and run statistics on. The field on the left specifies whether WebTrends will be accessing a local file or a file on a remote web or FTP server. The three choices are: file://, http://, and ftp://. You will recognize these as the prefixes you would type into a browser's Location box to access the resource from within a web browser.

The file type is pretty self-explanatory. Use this for a file on the local Windows computer. (The path field does provide a Browse button.) The http type is simple. If you've put your server log file in the document tree that your web server serves, you can access it through the server itself. The FTP type is also simple. However, you will have to supply the username and password that WebTrends will need to log into the remote web server to retrieve the file.

For example, say that your weekly script that rotates your log files places a copy of the current (which concurrently becomes last week's) log file into an anonymous FTP directory /var/spool/ftp/pub/logs/ and that the name of the log file is access.log. The URL you have to enter into WebTrends to access this file would be the following:

```
ftp.xyz.com/pub/logs/access.log
```

As soon as you change the file type field from the default file:// to ftp://, the Browse button at the right of the path field changes its label to Login. Click this button to enter the username and password for WebTrends to log into the FTP server (see Figure 9-3).

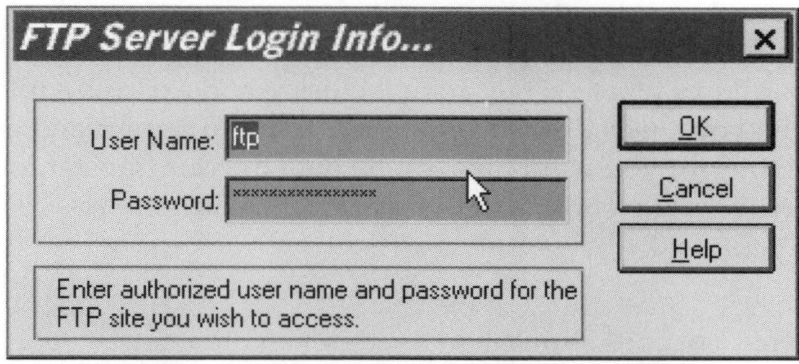

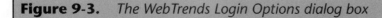

Figure 9-3. *The WebTrends Login Options dialog box*

If you use a Windows NT web server, such as Microsoft's own Internet Information Server or one of the Netscape servers for Windows, and you run WebTrends on the same system, you can take advantage of a neat feature that works only if the file type is file:// (which it would be, in this case). This feature is specifying filename wildcards and multiple filenames. This makes it convenient if your web server rotates logs (say, on a daily basis). For example, to import all of the files in the directory D:\web\logs, you could enter D:\web\logs*.* into the path box to cause WebTrends to import all of the log files in that directory.

Home Page Path

The final pair of fields in the Add/Edit Log File Entry dialog box specifies the top-level page of the web site for which you're running statistics. This is necessary for WebTrends to be able to access each page to get page titles—which it will do after the import is complete.

Import Options

By clicking the Filters button on the Add/Edit Log File Entry dialog box, you can set up import filters for the log file definition you are setting up. The default setting (Include everything) imposes no filters; it imports and processes the entire file. But if you want to report on, say, only visits from within the U.S. or only visits from Europe, you can set up a filter to do it.

In addition to filtering on geographic location (which is really just based on the domain name country code), you can set up filters based on domain types (.com, .edu., .mil, and so on), specific domains (such as Prodigy, CompuServe, or Netcom), regions of the U.S., and for specific return codes. For example, by default, WebTrends analyzes hits of all return values, but if you wanted to exclude requests that resulted in error status codes, here is where you would do it.

WebTrends allows you to set up both inclusive and exclusive filters. With inclusive filters, you specify just that data you want to import and analyze. Use exclusive filters if you want most of the data in your transfer log, but you want to eliminate certain subsets of that data from your reports.

Reports

WebTrends comes out of its shrink-wrap with seven predefined report templates:

- Executive summary
- Technical summary

- Top referring sites
- Complete summary
- Billing report
- Sites access by proxy report
- Quick summary

In addition to these reports, you can define your own. Here's how it works: WebTrends offers 44 statistical components. These are all of the detail-level components for every report—every statistic and graph that WebTrends knows how to run. These statistics include both tables and graphs for each of the following:

- General statistics
- User profile by regions
- Top requested pages
- Top downloaded files
- Top submitted forms and scripts
- Most active organizations
- Most active countries
- Activity statistics
- Activity by day of the week
- Activity by hour of the day
- Technical statistics
- Forms submitted by users
- Client errors
- Server errors
- Top download types
- Organization breakdown
- North American states and provinces
- Most active cities
- Most accessed directories
- Sites accessed by proxy

- Top referring sites
- Top referring URLs
- Most used browsers
- Most used platforms

The seven reports group some of these statistics together to form a focused report. For example, the Top Referring Sites Report includes only the top referring sites and URL graphs and tables. The Complete Summary, on the other hand, includes every statistical table and graph that WebTrends can produce.

Report Options

With WebTrends, you specify a date or time range restriction when you initiate a report. When you click on the Report toolbar button on the WebTrends' main program window, the Summary Report dialog box appears (see Figure 9-4). In this dialog box, you also choose your report template.

In the Report Range box, you can choose to report on an entire log or portions of the log, which are defined by a list of descriptive names; among the choices are First week of log, First day of log, and Last 6 months of log. Alternatively, you can use the Start Time, Start Date, End Time, and End Date boxes to specify the exact date and time ranges to include in your report.

Most often, you will report on an entire log that you have imported. This is especially true if you devise a log rotation scheme that rotates and archives your logs automatically. You have to click the Advanced button to access the remaining report options (shown in Figure 9-5). Here's where you can change the language of the reports. Besides English, WebTrends will happily produce your reports in French, German, or Japanese. You can also select one of your custom report templates and change the output file (the actual HTML filename) of the report.

In addition to changing the report filename, you can also configure WebTrends to FTP your log reports (and all of its associated graphics) to a remote machine. Using this feature, you can virtually automate the entire process. For example, if you make your log rotation script make the current (last) transfer log available in an FTP directory or in your web site's document tree, you can use a system scheduler (System Agent, At, or WinAt) to run WebTrends, which you will have configured to automatically run your reports at a preappointed time.

Summary Report...
Memorized Complete Summary
Start...
Cancel
Memorize...
Advanced...
Help
Report Range: All of Log
Start Time: N/A
Start Date: N/A
End Time: N/A
End Date: N/A

Figure 9-4. *The WebTrends Summary Report dialog box*

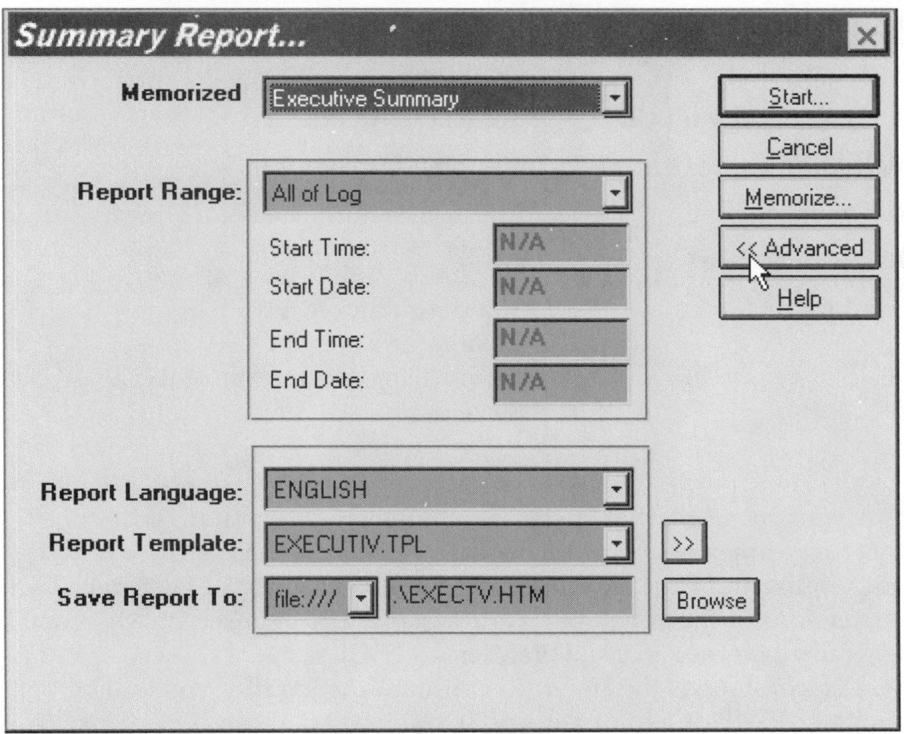

Figure 9-5. *Advanced options*

Summary

WebTrends is a solid log analysis package. It's easy to use, it's widely distributed, and it's priced competitively. While e.g. Software has some work cut out for it (the product can use some improvement in the quality of reports and graphs), WebTrends will be a major factor in this market.

Market Focus

Market Focus is the log analysis product of Intersé Corporation of Sunnyvale, California. Market Focus isn't quite as shrink-wrapped as its main competitor, WebTrends—and for good reason. Market Focus is a product with much more horsepower behind it and stronger legs beneath it. While WebTrends is a mass-market product targeting small- to medium-sized web sites, Market Focus is zeroed in on the medium- to large-sized web site.

Developer:	Intersé
Platforms:	Windows 95, Windows NT
Evaluation copy on CD?	Yes (standard edition)
Evaluation copies online?	Yes (standard edition)
Web site:	http://www.interse.com
Online purchase?	No
Retail price:	Standard edition, $695 Developer's edition (for Microsoft Access), $3,495 Developer's edition (for Microsoft SQL Server), $6,995

Three versions of Market Focus are available. The standard edition employs a run-time version of Microsoft's Access database as its internal database engine and runs on Windows 95 and Windows NT systems. There are two versions of the developer's edition: one is also based on Microsoft Access, while the other is based on Microsoft SQL Server.

Intersé recommends the standard edition if the log files you will be analyzing are 75MB or less in size and if you don't necessarily want to do anything fancy like complex custom analysis or aggregating site statistic data across multiple Internet sites. If you need any of this, you should look to the developer's edition. In addition, for high-volume, enterprise-scale operations and client-server functionality, you should look to the Microsoft SQL Server–based version for maximum scalability, flexibility, and robustness.

With the two Microsoft Access versions (the standard edition and the developer's Access edition), you don't need your own copy of Microsoft Access; a run-time version is included with the product. For the SQL Server–based developer's edition, you do need your own copy of Microsoft SQL Server.

Since the SQL Server version is based on a client-server database management system, you can run the Market Focus modules either on the system hosting the database management system itself or on other machines on your network. All you have to do is use the Microsoft ODBC subsystem on the client machines (the ones on which you will run the Market Focus software) to define the remote server system.

Import and Analysis Automation

Unlike WebTrends, which integrates the functions of import and report generation in a single application, Market Focus has two distinctly separate programs for data import and report generation. With WebTrends (targeting the lower-end market as it does), it is more important to have a single, simple-to-use interface. This is clearly a lower priority with Market Focus. In fact, since the operations of data retrieval and import are logically separate from the operation of running reports, it may even be desirable to have these as separate functions, as Market Focus does.

Both of the Market Focus modules (the Import and Analysis modules) have command-line interfaces. This makes prescheduling operations much more elegant than the WebTrends method. With command-line interfaces, you can use the built-in Windows scheduler to launch import or analysis operations.

To schedule import and analysis events on Windows NT systems, you need to be running the system Schedule service. (Start up the Schedule service in the Windows Control Panel if it isn't already running.) Then use the command-line program At or the WinAt program that comes with the NT Resource Kit to schedule your events. To schedule import and analysis events on Windows 95 systems, you'll need to pick up the Microsoft Plus! Pack if you don't already have it. Included with the Plus! Pack is a program called System Agent. System Agent is comparable in functionality to the Unix cron program, but it's a little easier to use. Figure 9-6 shows System Agent configured to run a batch file that will sequentially run the Import and Analysis modules.

The batch file you need to run differs a little bit depending on if you are using NT or Windows. With NT, regular batch commands will work fine, since the operating system waits for each command in a batch file to execute before moving on to the next. Here's a plain vanilla batch file for launching the Market Focus modules under Windows NT:

```
IMPORT.EXE db="interse.MDB" config=MySite.MFI log=June.LOG
ANALYSIS.EXE db="interse.MDB" config=Comprehensive.MFA report=June.DOC
```

In Windows 95, however, the operating system may move on to succeeding commands before the earlier commands have finished their work. The solution is to use the Windows Start command with the /Wait option to make Windows wait for each command to complete before starting the next one. Here's what the Windows 95 equivalent should look like:

```
START /Wait IMPORT.EXE db="interse.MDB" config=MySite.MFI log=June.LOG
START /Wait ANALYSIS.EXE db="interse.MDB" config=Comprehensive.MFA report=June.DOC
```

Under Windows 95, an automation scheme like this will work pretty much flawlessly. However, under Windows NT (at least pre-4.0), you have to make sure that the user under which the At job will run is logged in at the scheduled time of the run. This could throw some kinks into your plans if you hope both to have secure systems and to run your log analysis reports at odd times like Sunday mornings at 12:00 A.M.

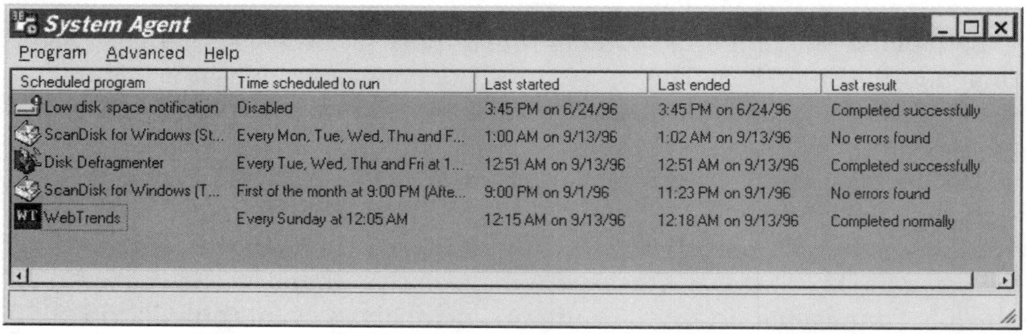

System Agent

Program Advanced Help

Scheduled program	Time scheduled to run	Last started	Last ended	Last result
Low disk space notification	Disabled	3:45 PM on 6/24/96	3:45 PM on 6/24/96	Completed successfully
ScanDisk for Windows (St...	Every Mon, Tue, Wed, Thu and F...	1:00 AM on 9/13/96	1:02 AM on 9/13/96	No errors found
Disk Defragmenter	Every Tue, Wed, Thu and Fri at 1...	12:51 AM on 9/13/96	12:51 AM on 9/13/96	Completed successfully
ScanDisk for Windows (T...	First of the month at 9:00 PM (Afte...	9:00 PM on 9/1/96	11:23 PM on 9/1/96	No errors found
WebTrends	Every Sunday at 12:05 AM	12:15 AM on 9/13/96	12:18 AM on 9/13/96	Completed normally

Figure 9-6. *The Windows 95 System Agent is configured to run a batch file, which in turn will execute the Market Focus Import and Analysis modules*

Data Import

Unlike WebTrends, which essentially imports your data and throws it away, Market Focus stores and keeps your log file data in its database until you purge it. This is a big advantage, because it makes rerunning reports and running reports with different criteria and time spans much quicker and simpler.

Unfortunately, as the program works now, the only criteria you can furnish to purge records from the database are the name of the site and a date before which every record will be deleted. This isn't always a bad thing. For example, let's say you actually want to purge records from your database because you want to trim the size of your data tables, and you really don't need the records anymore; the way you would do this is by providing the site name and a date before which every record will be deleted. However, if you want to delete an entire import (say, because you made a mistake and imported the wrong log file for a week or a month's period), then having only a "purge" capability can complicate your life a little.

Before you can actually import a log file, you have to set up your site in the Internet Site Manager (accessible on the Tools menu). Figure 9-7 shows Internet Site Manager with the fields filled in to define our cider mill site.

With your site defined, you are almost ready to import your server log file. Before you do, you may want to open the Import module options dialog box (see Figure 9-8) and configure Market Focus to look up the titles of HTML pages referred to in the transfer log; you do this by clicking the Lookup titles of new HTML files option. If you choose this option, Market Focus will use the titles of your pages in reports rather than just the path and filename of the HTML document itself. It's a small thing, but the small things can make a difference.

After you have set up your Internet site and import options, you return to the main window of the Import module (see Figure 9-9). Here you need to specify the local filename or remote FTP URL for your access log. If you use Market Focus to retrieve your log files from a remote FTP server, you will have to build the username and password into the URL as you would in an HTML link to an FTP server. In Figure 9-9, you can see the username (ftp) and the password (guest) as part of the FTP URL.

At this point, click the Start import button to begin the importation process. A status bar will appear with messages telling you what is going on, and a progress bar will give you an indication of how much longer the import will take.

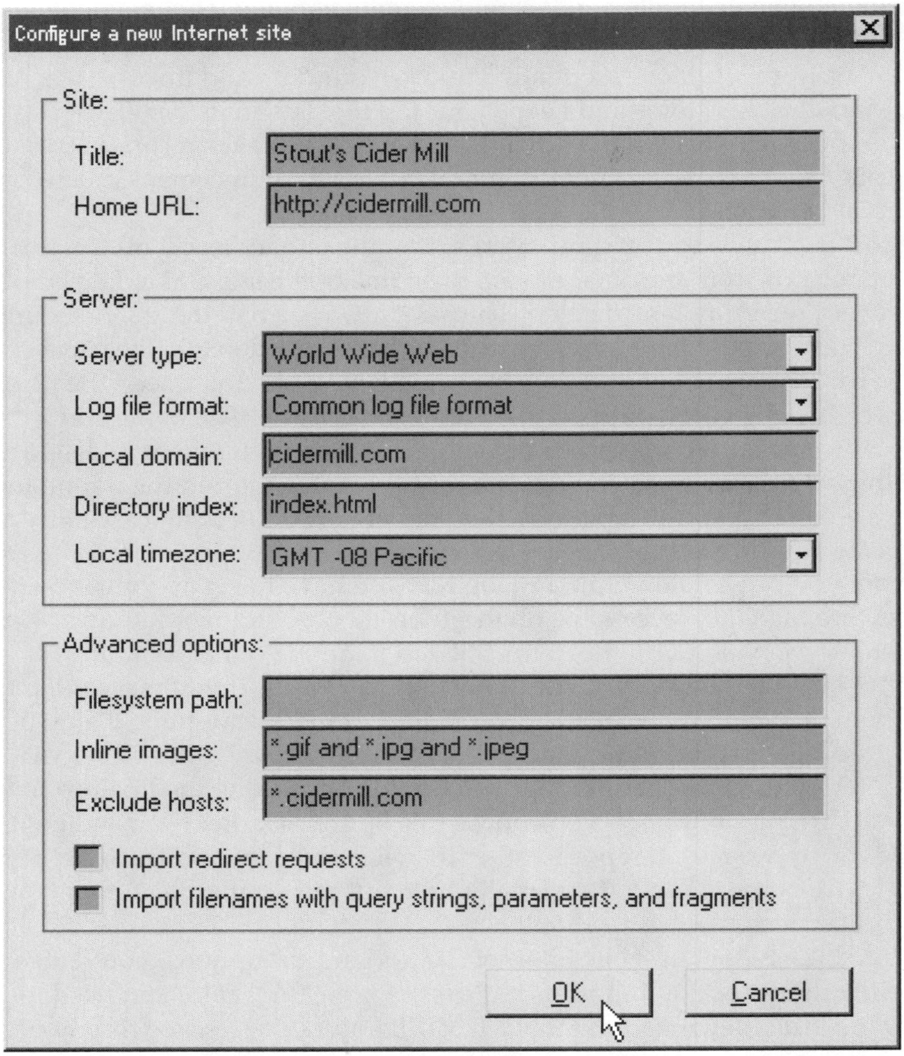

Figure 9-7. *Internet Site Manager*

Reports

The Market Focus Analysis module places you into a report editor of sorts, where you can choose from predefined reports or create your own custom reports. In Figure 9-10 is the Analysis module's main window, displaying a

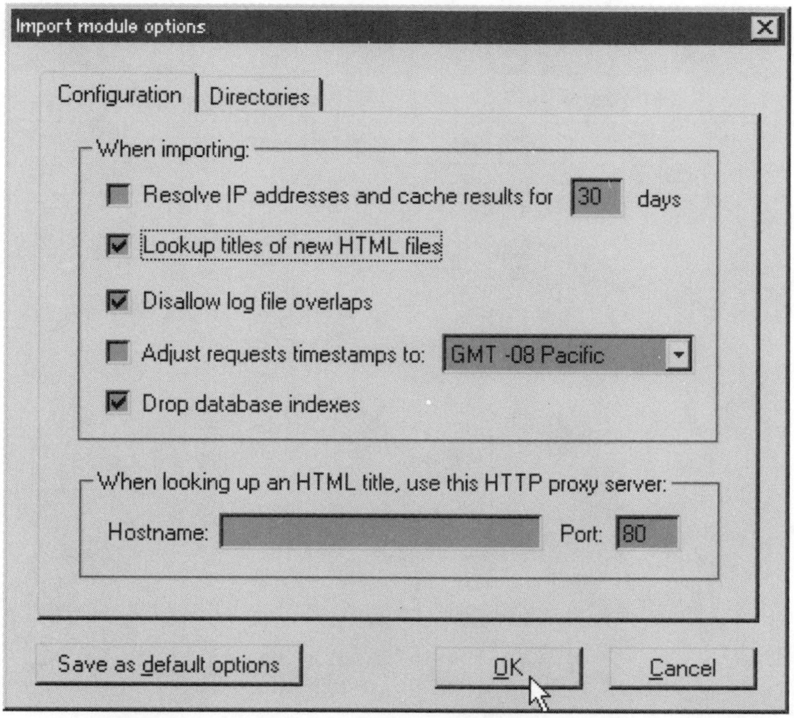

Figure 9-8. *The Market Focus Import module options dialog box*

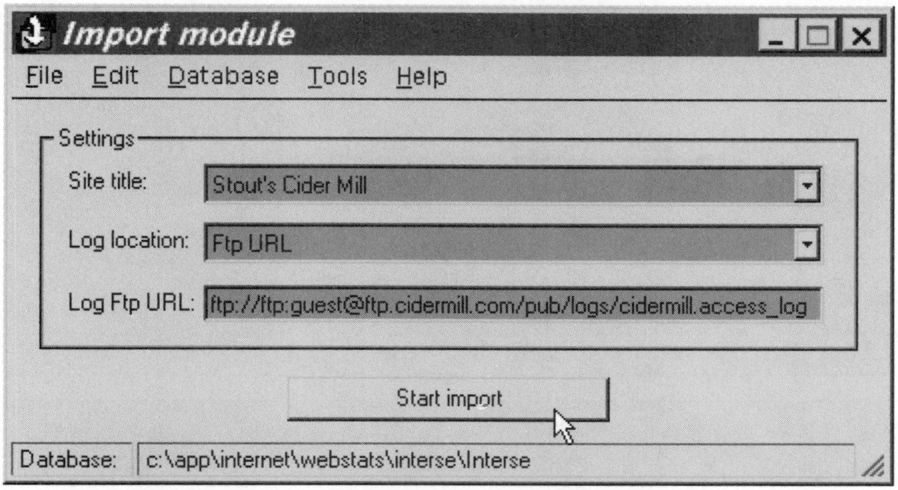

Figure 9-9. *The Market Focus Import module's main program window*

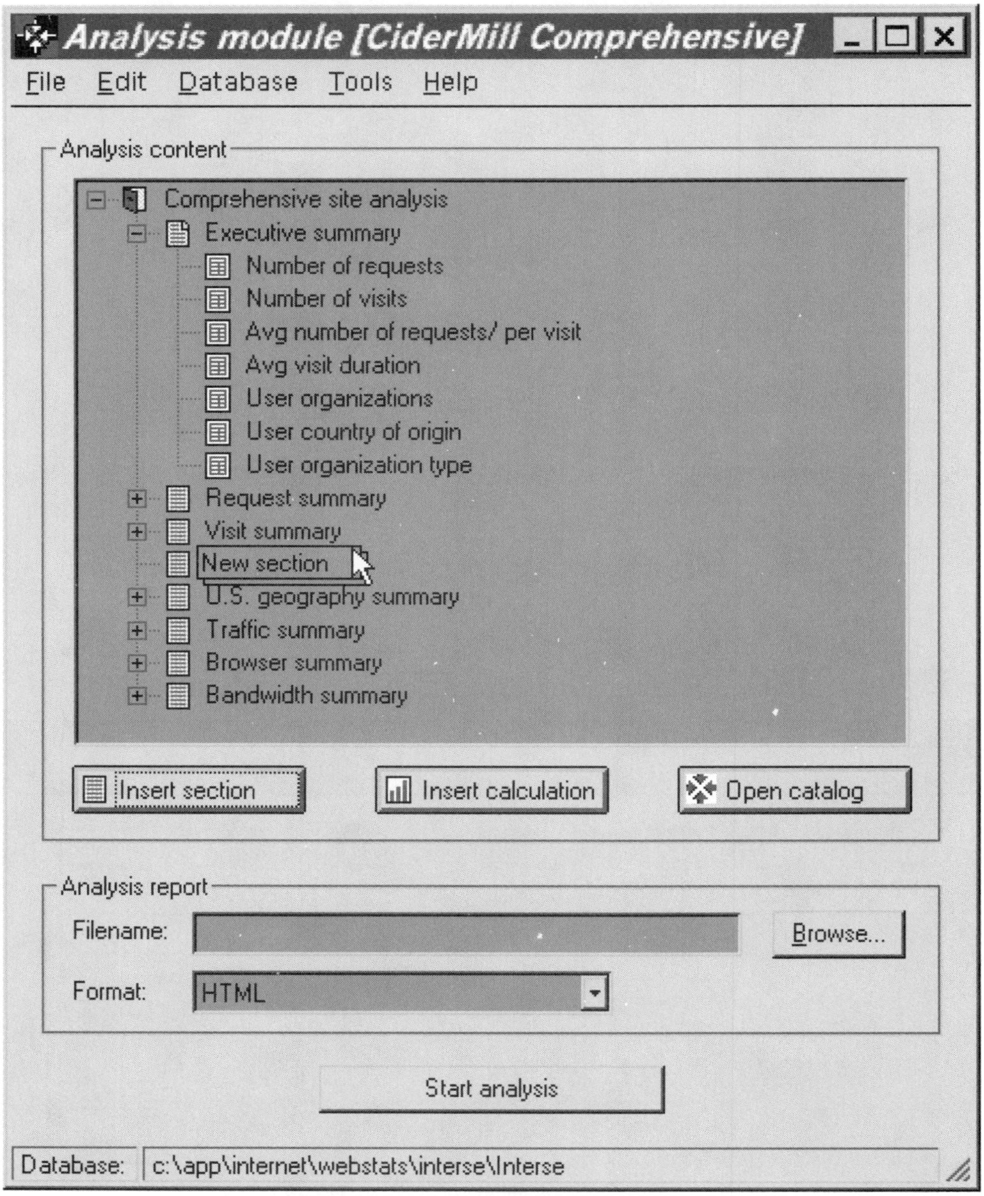

Figure 9-10. *Using the Market Focus Analysis module, you define your reports by manipulating an expandable/collapsible data tree*

previously defined and saved report format that was named CiderMill Comprehensive. The editor window displays the components (or sections) of your report in the order in which they will appear in the report. You can change the order of the components by dragging and dropping them. You can delete a component by right-clicking on it and selecting Delete from the pop-up menu. And you can add components by selecting a section and clicking the Insert section button. The new section will be inserted after the section that was highlighted when you clicked the button.

You can build your own report from scratch or start with one of the 14 predefined reports and refine it to meet your needs. After you insert a new section, you are prompted to define at least one new calculation for that section. The New calculation properties dialog box (see Figure 9-11) opens up,

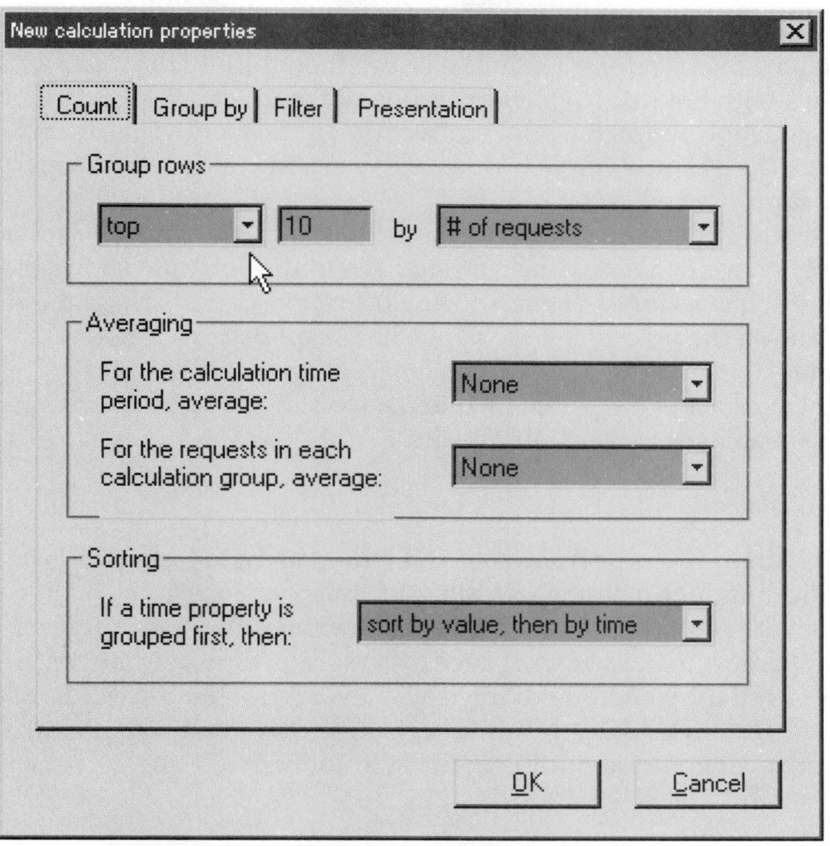

Figure 9-11. *The New calculation properties dialog box is where you define your own new statistical computations*

and you can define the details of your new computation including the content of the data (what it actually is). For example, you might define a computation to display a table of the least viewed pages in your web site. With such a definition, you can customize your new calculation by specifying the groupings and order of data, defining filters, and so on.

The Market Focus report format isn't flashy. It does include some generated graphics but nowhere near as many as WebTrends. This isn't necessarily a bad thing. After you run your statistics with WebTrends a few times, you may begin to wonder about the true usefulness of some of its graphs. The Market Focus reports eliminate the glitz—they don't even build in a background color or graphic. If those things are important to you, though, it's easy enough to add them into the HTML of the report yourself.

The statistic sections shine. Figure 9-12 shows the top of the standard Comprehensive site analysis. It clearly shows the site and the date ranges for which the report applies. Intersé has also thoughtfully provided a table of contents with links down to each section of the report. The Executive Summary section shows the total number of requests and visits during the period, the average number of requests per visit, and the average visit duration.

The Top documents report, shown in Figure 9-13, presents the ten most frequently viewed pages on the web site. The middle column shows the total number of requests for each page during the report period. The right-hand column shows the percentage each page had of all page requests.

Contrast this report to the Top entry pages report in Figure 9-14. This table shows the number of visits that included a view of the page question and the relative percentages for each of all visits.

Summary

Intersé's Market Focus package is an industrial-strength log analysis package for middle- and high-volume web sites. Its three versions give you the options that most web site administrators will feel they need for scalability and flexibility. Unfortunately, Market Focus also comes with an industrial-strength price tag. Even the standard edition is well over twice the price of its nearest competitor. But then again, this product isn't for the small web site owner who may be merely curious about the traffic at his or her web site.

net.Analysis and net.Analysis Desktop

net.Analysis and net.Analysis Desktop are products of net.Genesis Corporation of Cambridge, Massachusetts. Like Market Focus, net.Analysis

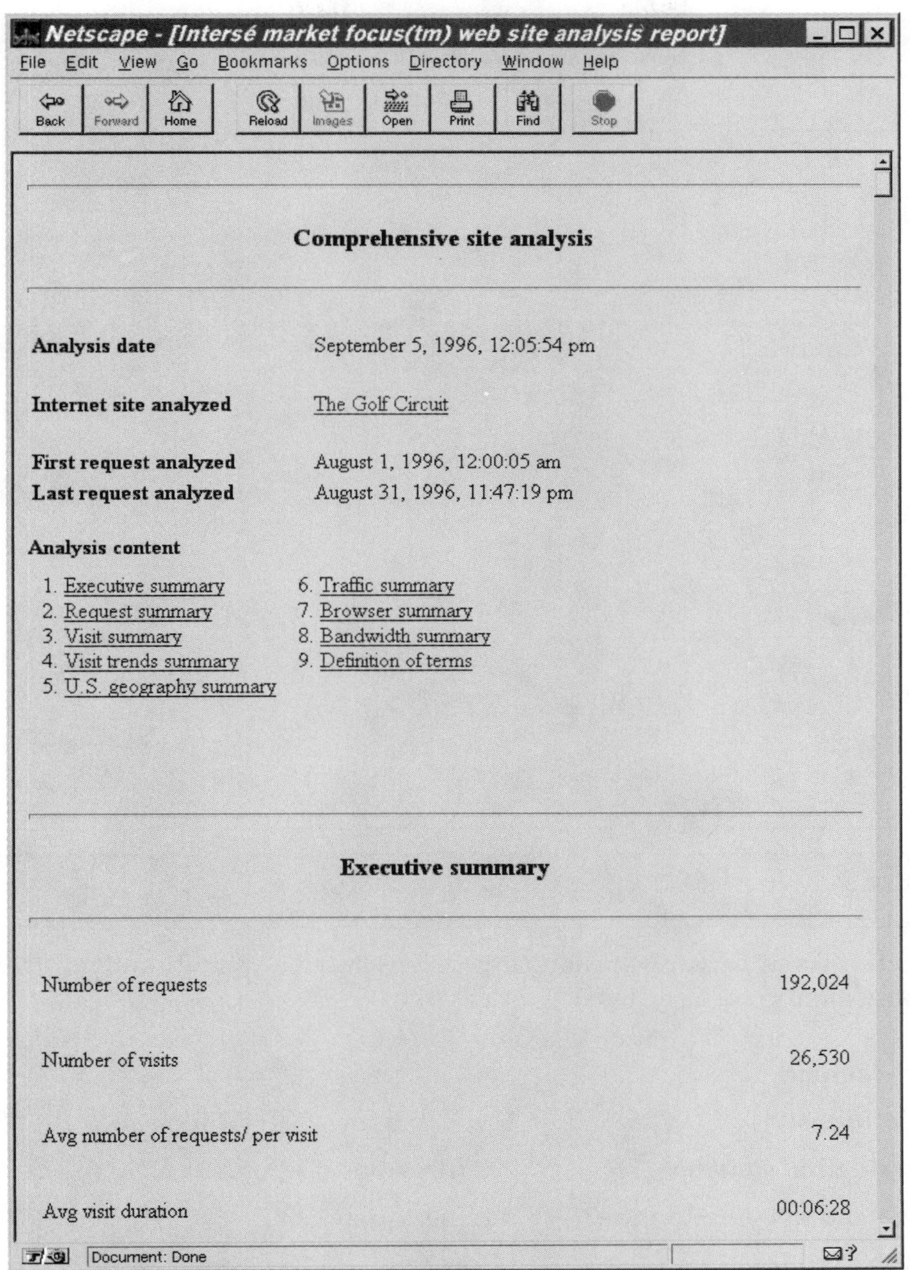

Figure 9-12. *The Market Focus Comprehensive site analysis report*

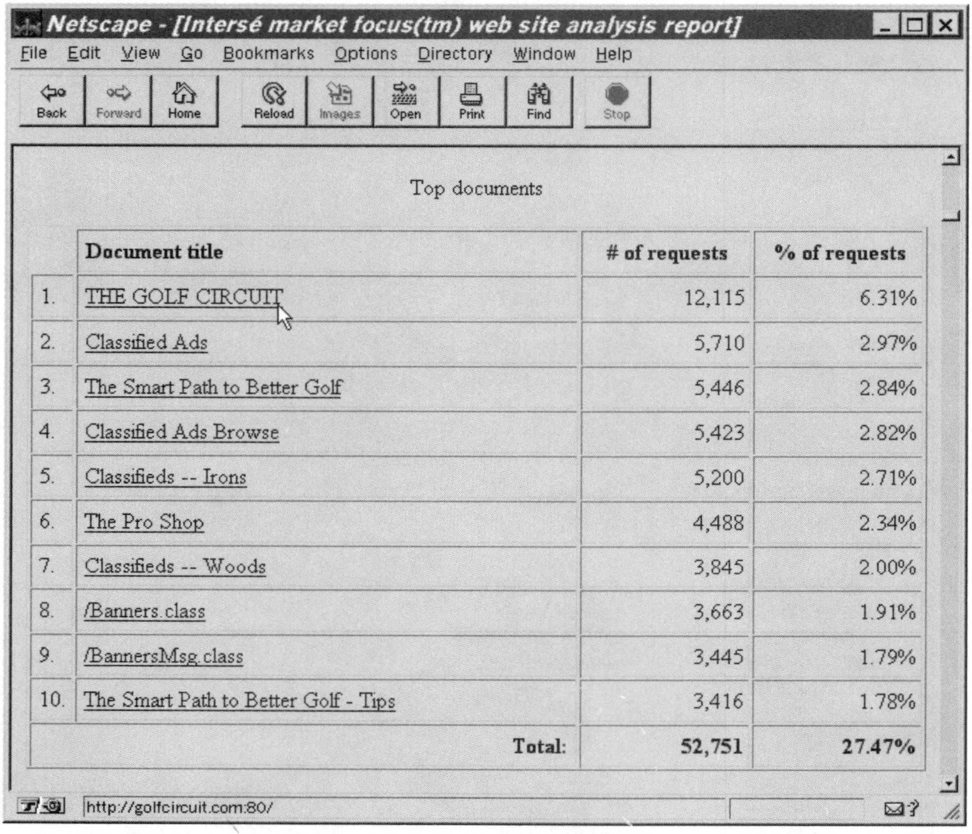

Netscape - [Intersé market focus(tm) web site analysis report]

Top documents

	Document title	# of requests	% of requests
1.	THE GOLF CIRCUIT	12,115	6.31%
2.	Classified Ads	5,710	2.97%
3.	The Smart Path to Better Golf	5,446	2.84%
4.	Classified Ads Browse	5,423	2.82%
5.	Classifieds -- Irons	5,200	2.71%
6.	The Pro Shop	4,488	2.34%
7.	Classifieds -- Woods	3,845	2.00%
8.	/Banners.class	3,663	1.91%
9.	/BannersMsg.class	3,445	1.79%
10.	The Smart Path to Better Golf - Tips	3,416	1.78%
	Total:	**52,751**	**27.47%**

http://golfcircuit.com:80/

Figure 9-13. *The Market Focus Top documents report*

and net.Analysis Desktop are similar products based on significantly different database management systems.

Developer:	net.Genesis
Platforms:	Unix, Windows NT, Windows 95
Evaluation copy on CD?	Yes (net.Analysis Desktop 1.1)
Evaluation copies online?	Yes (net.Analysis)
Web site:	http://www.netgen.com
Online purchase?	No
Retail price:	net.Analysis, $2,995 net.Analysis Desktop, $495

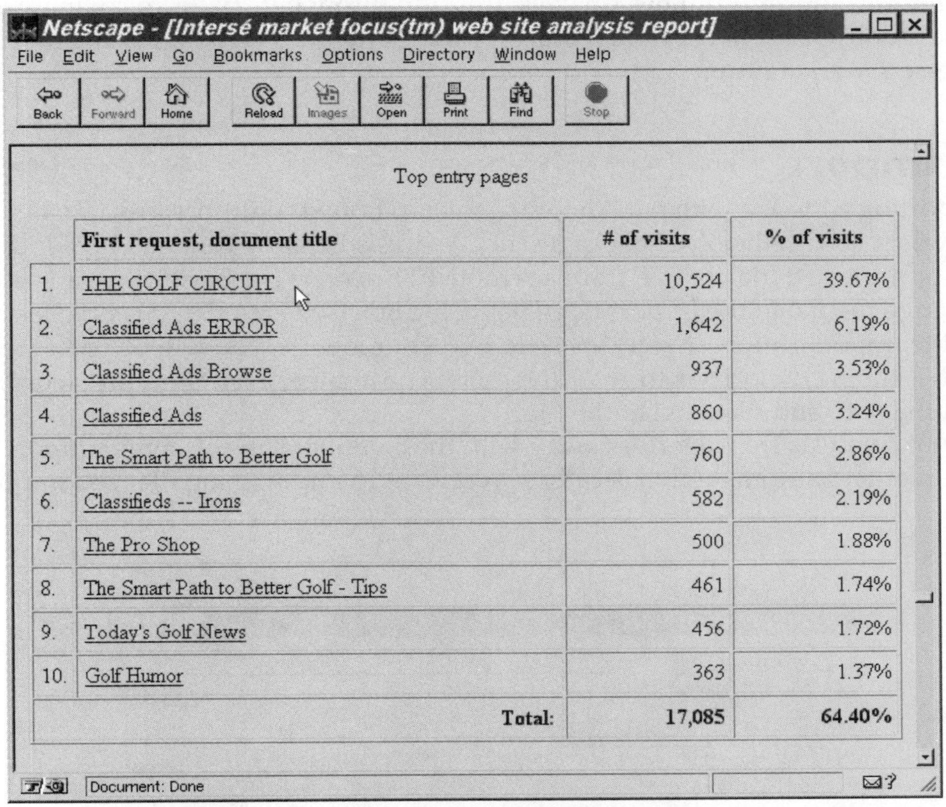

Netscape - [Intersé market focus(tm) web site analysis report]

File Edit View Go Bookmarks Options Directory Window Help

Back Forward Home Reload Images Open Print Find Stop

Top entry pages

	First request, document title	# of visits	% of visits
1.	THE GOLF CIRCUIT	10,524	39.67%
2.	Classified Ads ERROR	1,642	6.19%
3.	Classified Ads Browse	937	3.53%
4.	Classified Ads	860	3.24%
5.	The Smart Path to Better Golf	760	2.86%
6.	Classifieds -- Irons	582	2.19%
7.	The Pro Shop	500	1.88%
8.	The Smart Path to Better Golf - Tips	461	1.74%
9.	Today's Golf News	456	1.72%
10.	Golf Humor	363	1.37%
	Total:	**17,085**	**64.40%**

Document: Done

Figure 9-14. *The Market Focus Top entry pages report*

net.Analysis is a client-server-based system composed of a server-side database engine (Informix) and a client-side Reporter. As I write this, the database engine is available only for the Sun Solaris Unix platform, but net.Genesis promises to have it available for other large-scale Unix operating systems (including AIX, HP-UX, IRIX, OSF/1, and SunOS) in the near future.

The net.Analysis Reporter module (the client side) is available for Windows 95 and Windows NT. net.Genesis promises a Unix client in the near future.

net.Analysis Desktop is a completely different product. It is a stand-alone application with its own integral database. net.Genesis chose Microsoft FoxPro as the database engine for the Desktop edition.

For this review, I put net.Analysis Desktop through its paces to compare its functionality, ease of use, and features to e.g. Software's WebTrends and Intersé's Market Focus.

Data Import

Importing a log file with net.Analysis Desktop is as straightforward and as easy as it is with the other packages. net.Analysis Desktop can load up a local file or retrieve a file via FTP from a remote FTP server. The basic process is to describe the file in the Import Log File dialog box (see Figure 9-15) and click on the Import button. If you choose to FTP your log file, you will be presented with a simple dialog box to enter the remote hostname, username, password, and the path and filename of the file.

net.Analysis Desktop does away with the complications of saving import schemes and automation. Basically, if you want to import a log file, you import

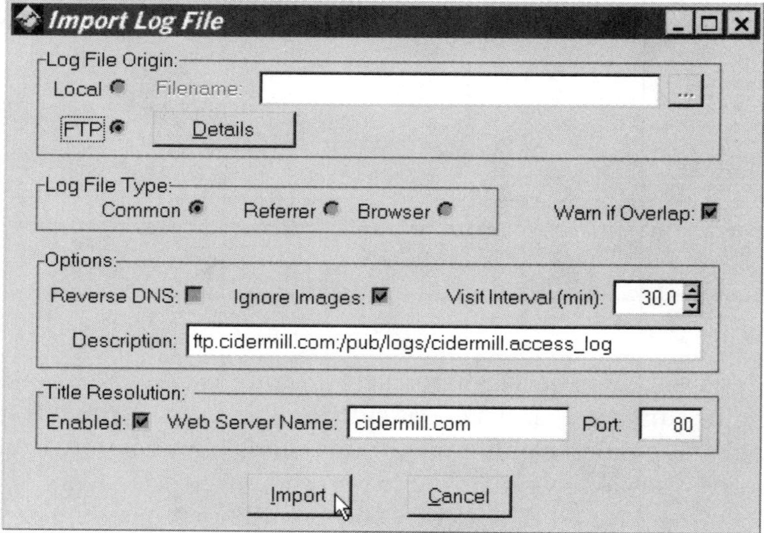

Figure 9-15. *The net.Analysis Desktop Import Log File dialog box*

it. Once it is done, your import will show up in net.Analysis Desktop's Log File Manager, shown in Figure 9-16. The Log File Manager keeps a record of every import and deletion you have done for each database you keep.

Reports

While importing log files is similar to the other products, that's about where any similarities with the other products end. net.Analysis Desktop's reports are fundamentally different in that the primary objective is to display its reports on the screen rather than create an HTML document with tables and graphs. Also, rather than running a single report to produce all of the statistics, graphs, and tables at once, with net.Analysis Desktop, you run one report at a time. You can view the report and its accompanying graph (if applicable) and choose to save it to an HTML file, but this isn't the primary objective of net.Analysis Desktop.

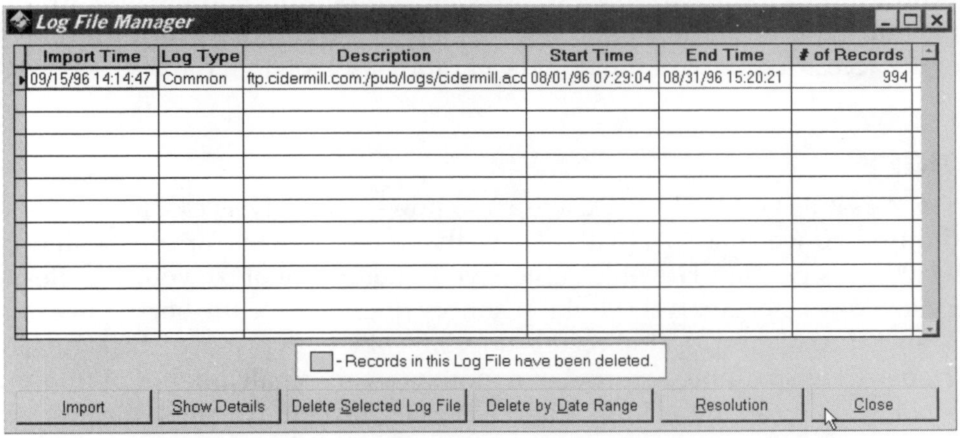

Figure 9-16. *The net.Analysis Desktop Log File Manager*

net.Analysis Desktop reports include the following:

- Visits per time
- Hits per time
- Bytes per time
- Top domains
- Top subdomains
- Top hostnames
- Top domain classes
- Top resources by filename
- Top resources by title
- Top resource classes
- Top referrals
- Top browsers
- Error analysis

Figure 9-17 shows the Visits Per Time report for the cider mill site. When you run a report, net.Analysis Desktop opens a program child window to display the report output. On the left is a table of the summary data process from the access log file. On the right is a graph based on that data.

Summary

net.Genesis bills net.Analysis Desktop as a powerful analysis tool for small- to medium-sized web sites. Actually, it says that the product is for sites with up to 100,000 hits per day. However, unless you intend to analyze your log files every few days, net.Analysis Desktop simply doesn't have the horsepower.

I ran net.Analysis Desktop on the cider mill site with just over 2,000 hits in the transfer log spanning a period of a month (a very small amount of traffic), and it breezed through the import and reports. I also ran it against the golf site, with nearly a million hits in the transfer log spanning the same month. net.Analysis Desktop simply couldn't handle that kind of volume—at least not gracefully (note that one million hits over a period of a month is only around 33,000 hits per day). So, clearly, the truth of the 100,000 hits per day claim only applies if you import and rotate your logs often.

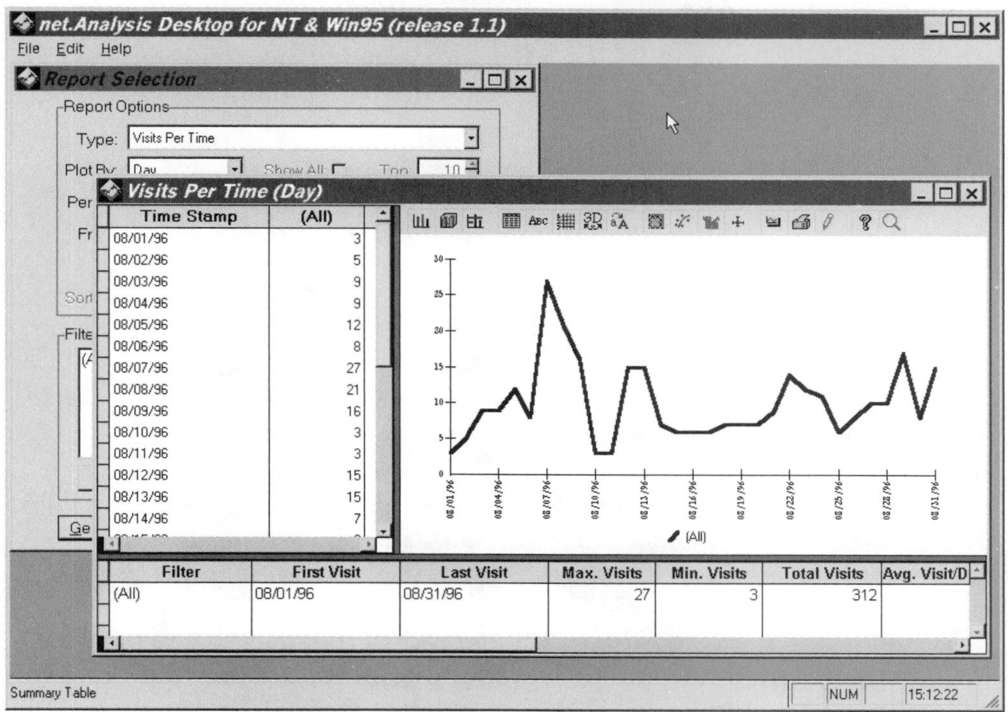

Figure 9-17. *The net.Analysis Desktop Visits Per Time report appears in a program child window; you can, however, export it to an external HTML file format*

Although the reporting methodology is different and interesting, it is a little more cumbersome if you want your reports in HTML. In that case, you have to run each report separately, export each of them to HTML files, then manually create a page that links them together or otherwise joins them.

Other Commercial Software

Besides the three packages we have looked at in this chapter there are other commercial log analysis software packages you may want to look into. In this chapter, we have explored the three main packages that operate under Windows 95 and Windows NT. The remaining three (of the most important) packages are Unix-only software or work only with a subset of web servers.

SiteTrack from Group Cortex is one of these products. Group Cortex kindly contributed binary copies of their software for several Unix servers to the CD at the back of this book. These Unix versions are for the following systems:

- IBM AIX 3
- IBM AIX 4
- HP-UX 9
- IRIX 5
- SunOS 4
- SunOS 5

While these versions of SiteTrack are on the CD, you will need a Unix computer (of one of the above persuasions) to be able to use them. Also, SiteTrack doesn't work with just any web server that produces Common Log Format or combined (extended) log format log files; it works only with Netscape servers. So if you use one of these Unix systems—and you use a Netscape web server—check out SiteTrack.

Open Market's WebReporter is another product that falls somewhat into this category. (Open Market didn't supply any of its software for the CD in this book, but it did provide a nice HTML presentation about its product.) WebReporter runs on a variety of Unix systems, including the following:

- Sun Solaris
- SunOS
- IBM AIX
- BSDI
- HP-UX
- SGI
- Digital UNIX

WebReporter takes a completely different approach to log analysis than most of the other packages. WebReporter is actually a separate server that you install on the system that runs your web server. It works like a web server itself: it serves up its interface through HTML. Using this interface, you can define, run, and view the reports that you want to see on your web site traffic—all through an interactive web interface.

Lilypad from Streams is another Unix-only system. Lilypad also offers an interactive HTML interface for running reports. Streams has built into its system some innovative report features, such as comparative reports (this period compared to last period; the change in real numbers and as a percentage).

Streams also did not provide any software for this book (either for the CD or for review). They did, however, contribute a portion of their online web site—complete with sample interfaces and reports for the CD-ROM.

In this chapter, we've looked at several of the most popular commercial log analysis software packages. We've looked at some more closely than others. In the end, you will want to download and evaluate each of the packages in which you are interested and decide for yourself which ones suit your needs. In the meantime, you have a good idea of how several of them work. In Chapter 10, we will take a similar look at traffic tracking service companies and how their reports and analyses differ from those of log analysis software packages that you purchase and run yourself.

COLOR
FILM

Chapter Ten

Traffic Tracking Services

For some folks, doing it yourself just isn't worth the effort. There's a lot to be said for this approach—especially when you're talking about highly technical subjects, like analyzing web server log files, and you would rather be spending your time polishing and fine-tuning your site's content. Two major players in the log analysis service field make it easy to leave the driving to them.

These players are NetCount, LLC and I/PRO Corporation, and both of them offer traffic tracking services. Your web server continues to work just the way it always has, but the log file data gets sent to the service company for processing. They continually upgrade and maintain their own log analysis software, which they run on your log data.

The only software you have to install on your servers is a small program that gets launched periodically out of your server's crontab file. This program sends the latest log data to the service company. It takes up little disk space, puts a very small load on your CPUs, and takes virtually no time or intervention on your part to keep the whole process running smoothly. When you want to view your site traffic reports, you just go to a sign-in page on your particular service company's web site, log in with your site URL and a password, and peruse your reports.

In this chapter, we will look at the services of one of these companies (NetCount) in great detail. We will explore the mechanics of setting up a system to use the NetCount service. We will also explore NetCount's reports and terminology and see how these reports and features differ from those you get by running your own log analysis software. In addition, this chapter will briefly cover I/PRO services.

NetCount, LLC

Located in Los Angeles, NetCount is a fast-growing small company with an excellent service. With only a phone call, NetCount can have you up on their service in just a few hours (although it does take a few days to get into the report queue). Once you're up, NetCount will automatically generate your reports nearly every day.

While the NetCount service isn't exactly cheap compared to buying your own log analysis software outright, it won't break a small budget either. As I write this, NetCount's Basic service is going for $98 per month for sites with fewer than 200,000 hits per day. They consider any site with fewer than 200,000 hits per day a startup site. (In fact, by NetCount's scale, a "small" site gets between 200,000 and 400,000 hits per day.) The rates for both the Basic plan and the Plus plan increment upward according to the average daily number of hits your site draws. See the web site at http://w w w.netcount.com for the latest rate information.

The difference between the Basic service and the Plus service is the number of reports you get. With the Basic service, you get the following reports:

- Distinct Point of Origin (DPO) Report
- Page Information Transfer (PIT) Report
- Abandonment Report

With the Plus service, you get the three Basic reports plus the following:

- Top Site Pages Report
- BIT/PIT Comparison Report
- File Transfer by Type Report
- Root Domain Report
- Top Ten Organizations by Sub Domain Report
- Error and Abandonment Report

For this evaluation, NetCount graciously provided an account (the Plus service) for the golf site. In the following sections, we will look at each of the reports that NetCount produces.

Installation

Installing the NetCount software is a snap. It supports every major flavor of Unix systems, including SunOS, IRIX, Solaris, FreeBSD, BSDI, AIX, HPUX, Linux, UNIXware, and DEC OSF/1. In addition, NetCount also works with servers running under Windows NT.

As for log file formats, NetCount supports the Common Log Format, the combined or extended log format (which includes referrer and user agent information), and the Microsoft IIS log format (the log files produced by Microsoft Internet Information Server).

The software that you actually install on your own server is minimal. All the software really does is take the latest data from your transfer log, compress it, and send it off to NetCount via a secure connection. You do have to make an entry in your system's crontab table to run the software every hour. If you use a Windows NT system, you would use the system scheduler (or comparable At program) to schedule transfers.

Depending on your particular system and web server, NetCount will supply you with an appropriate installation package. For Unix systems, this will most likely be a gzipped tar file for which NetCount will give you an FTP URL to retrieve. Once you retrieve this file and unarchive it, all you have to do is run the installation script which is tailored to your system.

Running the installation script is quick and simple. You just answer a few questions about where you want your NetCount files to reside, the location of your server transfer log, and whether or not you want the script to put the entry in your crontab file to execute the transfer program automatically. That's all there is to it. Within an hour of installation, your server will be automatically sending your log data to NetCount for processing.

NetCount Terminology

For one reason or another, NetCount decided to invent their own terminology with respect to web site statistics. The terms are a little confusing at first, but perhaps in the long run they will help to standardize and clarify industry-wide terminology. Table 10-1 presents each of the terms that NetCount uses—and they use them in all of their reports, so you might as well become familiar with them.

Term	Description
BIR	BIR stands for browser information request. To the rest of us, this is a regular "hit" (or request) in the transfer log. BIRs can be requests for HTML pages, graphics, sound objects, video, or whatever.
BIT	BIT stands for browser information transfer. These are requests, or hits (or BIRs), that result in a status 200 (OK) or status 304 (Not changed). In other words, it's the result of a successful request.
PIR	A PIR (for page information request) is a request (or BIR) for an HTML page. This is basically the BIRs less those requests that were for graphics, video, and so on.

Table 10-1. *NetCount Terms*

Term	Description
PIT	A PIT (for page information transfer) is a successful PIR (a request for an HTML document that resulted in success).
BIE	BIE stands for browser information error. This is a request (for any object) from a browser that results in something less than success. This would include broken links to images, requests for pages that don't exist, and so on.
BTA	A BTA, or browser transfer abandonment, may be the result of a reader clicking on the Stop button prior to a page loading completely.
DPO	DPO stands for distinct point of origin. This is analogous to a visitor with other software. However, NetCount's DPO doesn't take anything special into account to identify unique visitors. The DPO is based solely on the domain name or IP address of visitors.

Table 10-1. *NetCount Terms* (continued)

Regarding NetCount's use of the DPO: while they use only the domain name or IP address to identify what the rest of us call visitors, they do offer an upgraded service called HeadCount. This service employs a more wholesome algorithm to determine visitors, taking into account factors like multiuser computer systems and cookies.

NetCount Daily Reports

To view your NetCount reports, you drop into the NetCount web site and click on the Customer Report Pick-up button on the header frame of any page. This brings you to NetCount's sign-in screen, shown in Figure 10-1, where you enter your company URL and your preassigned password.

Your main report interface is a calendar of the current month (see Figure 10-2). To access a daily report, click on the link in the calendar for the day in which you are interested. For reports from previous months, you can click on the Monthly Menu link above the calendar. This will present you with a list of links to the previous months for which you have generated reports.

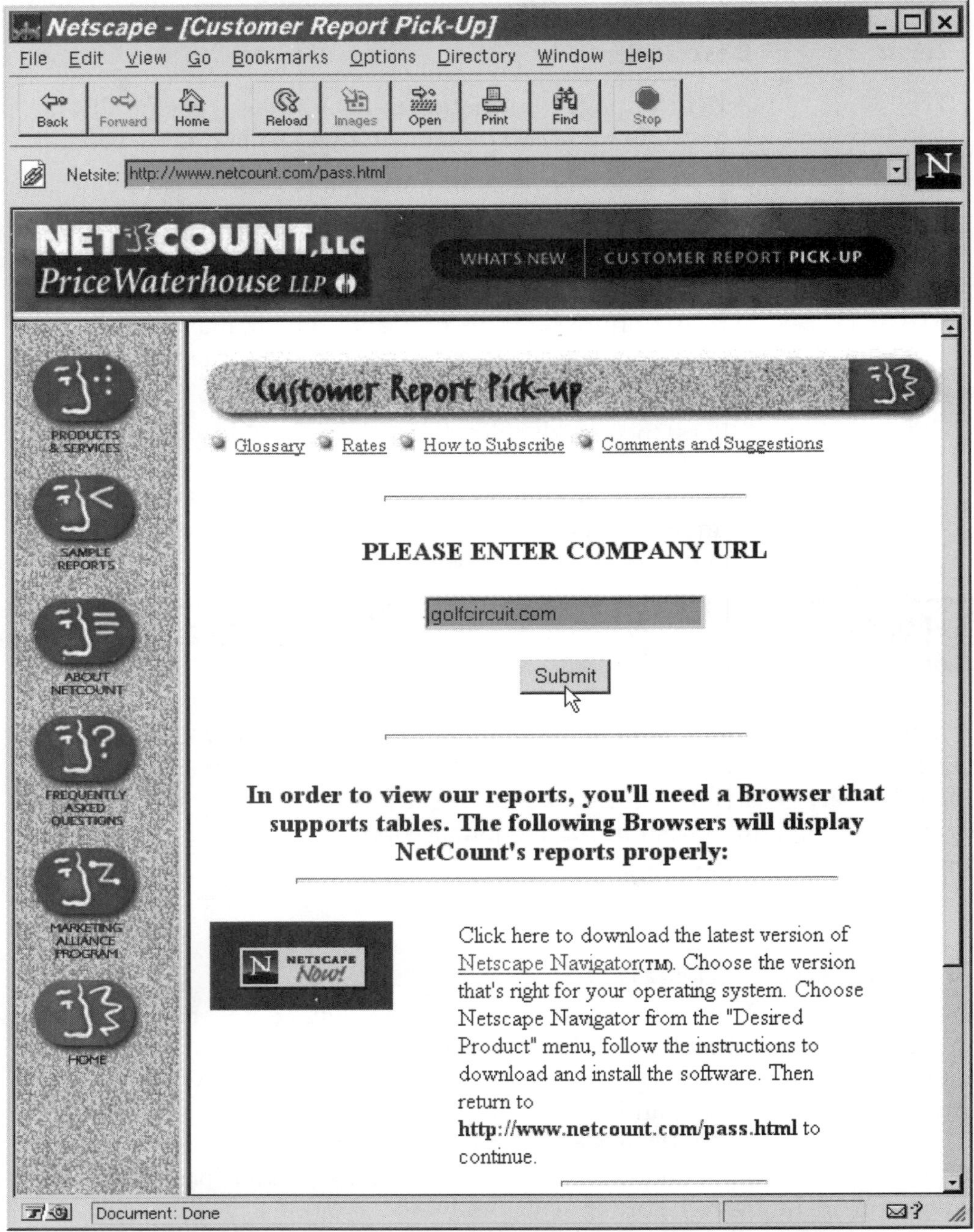

Figure 10-1. *The NetCount report sign-in screen*

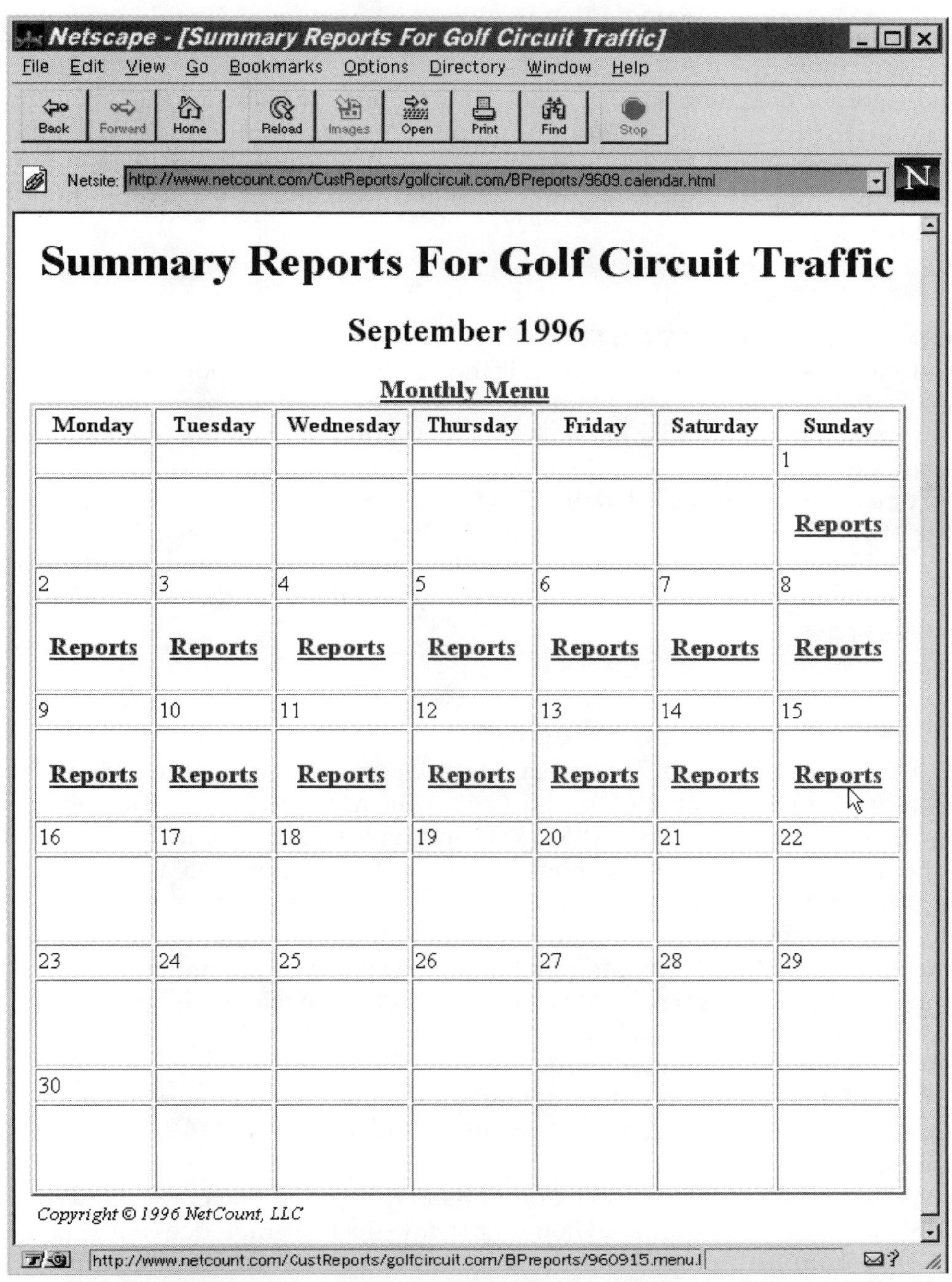
Netscape - [Summary Reports For Golf Circuit Traffic]
File Edit View Go Bookmarks Options Directory Window Help
Back Forward Home Reload Images Open Print Find Stop
Netsite: http://www.netcount.com/CustReports/golfcircuit.com/BPreports/9609.calendar.html
Summary Reports For Golf Circuit Traffic
September 1996
Monthly Menu
Monday Tuesday Wednesday Thursday Friday Saturday Sunday
1 Reports
2 3 4 5 6 7 8
Reports Reports Reports Reports Reports Reports Reports
9 10 11 12 13 14 15
Reports Reports Reports Reports Reports Reports Reports
16 17 18 19 20 21 22
23 24 25 26 27 28 29
30
Copyright © 1996 NetCount, LLC
http://www.netcount.com/CustReports/golfcircuit.com/BPreports/960915.menu.l

Figure 10-2. *The Summary Reports calendar*

The links for every day in the calendar (except for Sundays) lead to your report for the particular day. The links in the Sunday boxes lead to both daily reports for the particular Sunday and weekly reports summarizing the previous week's activity. Figure 10-3 shows the Report Menu screen under the link for Sunday, September 15. Notice the weekly reports listed below the daily reports. The equivalent pages under any other day look similar—they just omit the weekly reports.

Top Site Pages Report

The first of NetCount's daily reports is the Top Site Pages Report. Shown in Figure 10-4, this report shows you the top ten viewed pages on your web site. It's arranged in descending order of PITs and shows the abandonment rate for each page.

For each of the pages listed, the Top Site Pages Report also shows the peak hour of views (and the number of views in that hour), the average time spent on each page, and a small line graph displaying the activity for that particular page throughout the day juxtaposed against the activity for that same page the previous day.

BIT/PIT Comparison Report

The BIT/PIT Comparison Report, shown in Figure 10-5, presents a table with a row representing each hour of the day (from 12:00 A.M. to 11:00 P.M.). The columns line up the BIRs/BITs and the PIRs/PITs and report a success rate for each.

Looking at the top row (the hour of 12:00 A.M.) in Figure 10-5, you can see 352 BIRs (or hits). Of that number of requests, 344 of them were successful, for a success ratio of 97.7 percent. However, of those total browser requests, only 66 were for actual HTML pages (the PIRs), and all 66 page requests resulted in success, for a success rate of 100 percent. This tells you that the 2.3 percent of BIRs that were not completely transferred were all non-HTML pages (and were probably inline graphic images).

Scanning down the report page, you can see that the success rate is fairly consistently above 95 percent, both for BIRs (all requests) and PIRs (page requests).

At the bottom of the BIT/PIT Comparison Report is a line graph showing the BIT and PIT activity for each hour of the day (the line graph does not appear in Figure 10-5). This graph can help you quickly spot the peak activity times during the day for both BITs and PITs.

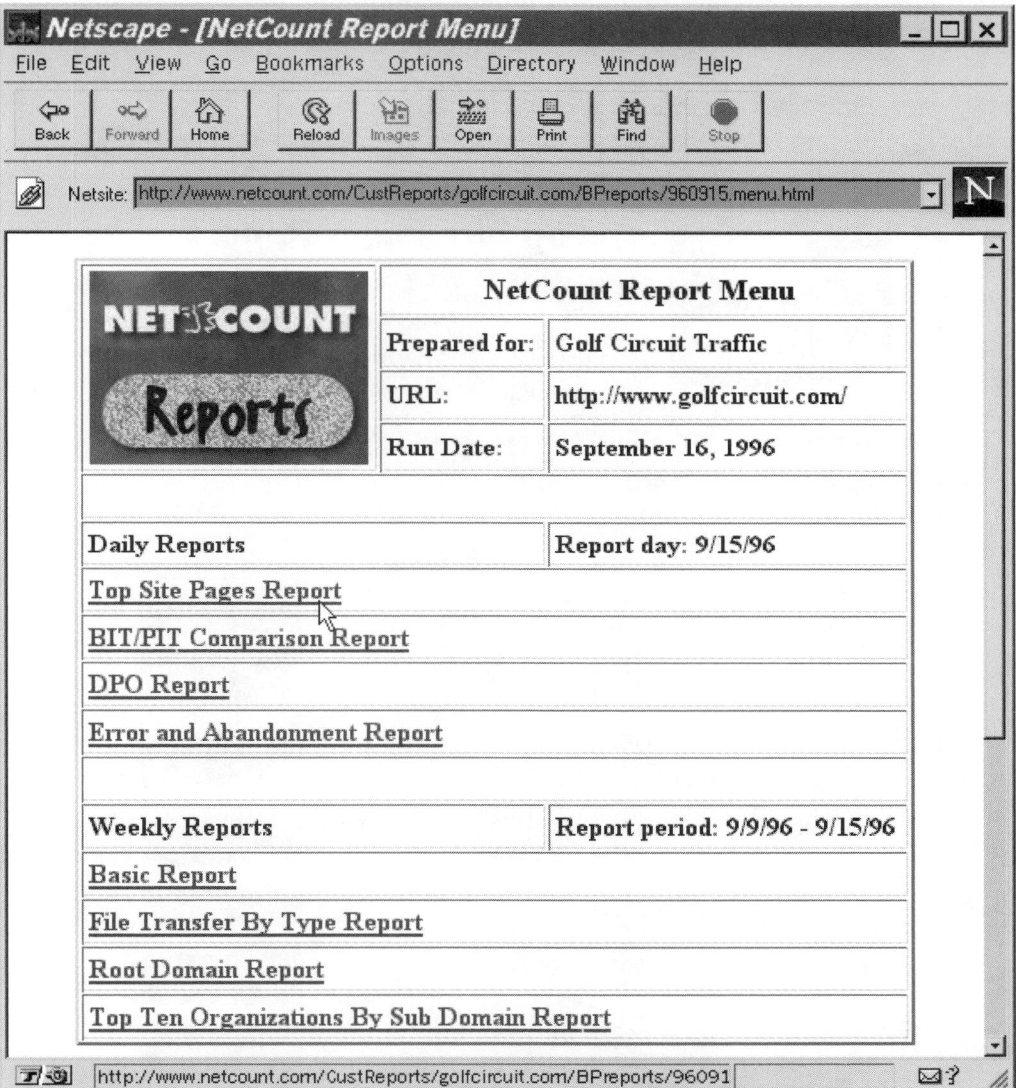

Figure 10-3. *The Report Menu, showing daily and weekly reports*

Distinct Point of Origin (DPO) Report

The Distinct Point of Origin Report (or DPO Report) shows the total number of unique visitors that visited your site during each hour of the day. This report is

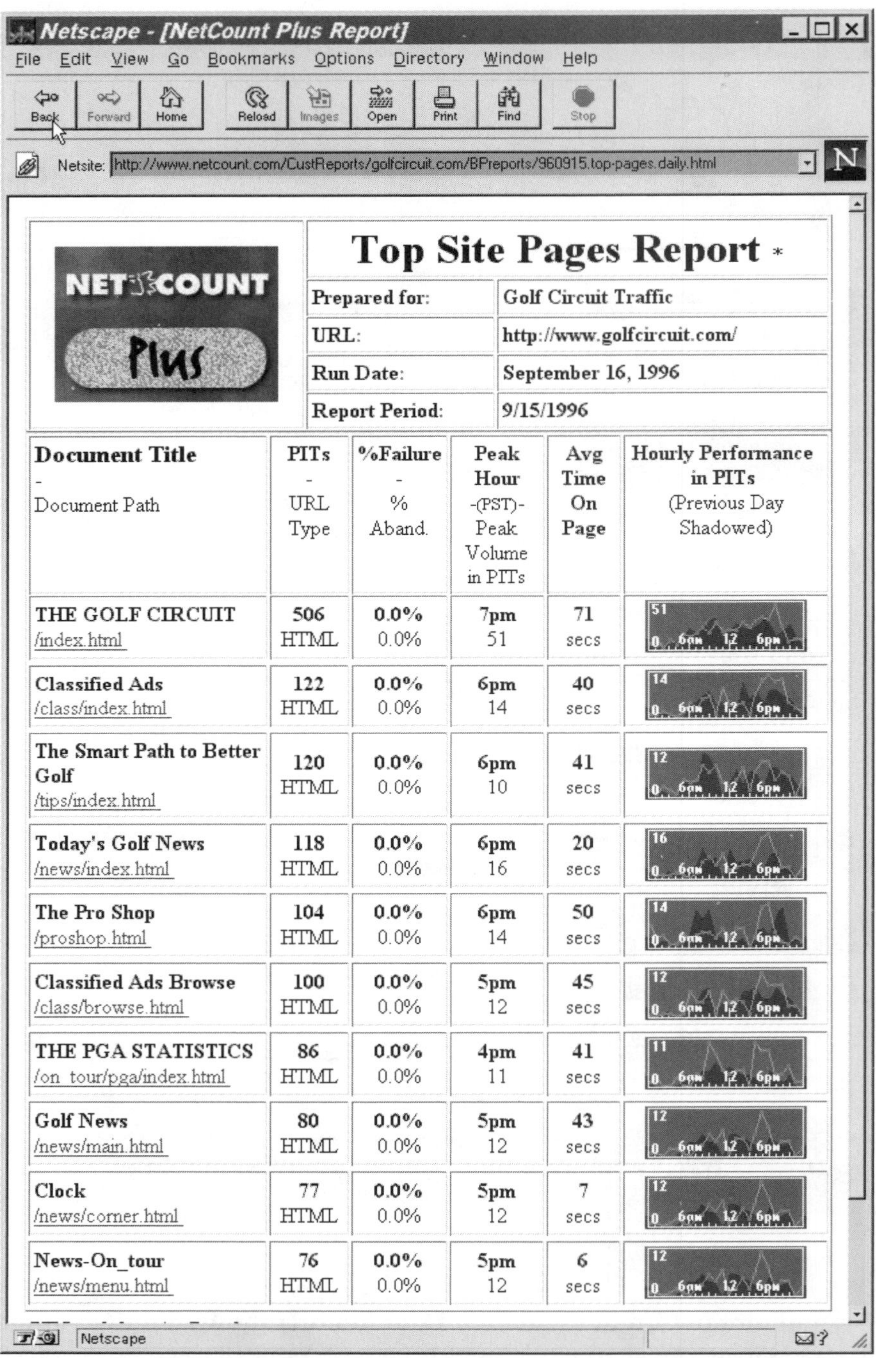

Top Site Pages Report *

Prepared for:	**Golf Circuit Traffic**
URL:	**http://www.golfcircuit.com/**
Run Date:	**September 16, 1996**
Report Period:	**9/15/1996**

Document Title - Document Path	**PITs** - URL Type	**%Failure** - % Aband.	**Peak Hour** -(PST)- Peak Volume in PITs	**Avg Time On Page**	**Hourly Performance in PITs** (Previous Day Shadowed)
THE GOLF CIRCUIT /index.html	**506** HTML	**0.0%** 0.0%	**7pm** 51	**71** secs	51 0 6am 12 6pm
Classified Ads /class/index.html	**122** HTML	**0.0%** 0.0%	**6pm** 14	**40** secs	14 0 6am 12 6pm
The Smart Path to Better Golf /tips/index.html	**120** HTML	**0.0%** 0.0%	**6pm** 10	**41** secs	12 0 6am 12 6pm
Today's Golf News /news/index.html	**118** HTML	**0.0%** 0.0%	**6pm** 16	**20** secs	16 0 6am 12 6pm
The Pro Shop /proshop.html	**104** HTML	**0.0%** 0.0%	**6pm** 14	**50** secs	14 0 6am 12 6pm
Classified Ads Browse /class/browse.html	**100** HTML	**0.0%** 0.0%	**5pm** 12	**45** secs	12 0 6am 12 6pm
THE PGA STATISTICS /on_tour/pga/index.html	**86** HTML	**0.0%** 0.0%	**4pm** 11	**41** secs	11 0 6am 12 6pm
Golf News /news/main.html	**80** HTML	**0.0%** 0.0%	**5pm** 12	**43** secs	12 0 6am 12 6pm
Clock /news/corner.html	**77** HTML	**0.0%** 0.0%	**5pm** 12	**7** secs	12 0 6am 12 6pm
News-On_tour /news/menu.html	**76** HTML	**0.0%** 0.0%	**5pm** 12	**6** secs	12 0 6am 12 6pm

Figure 10-4. *The Top Site Pages Report*

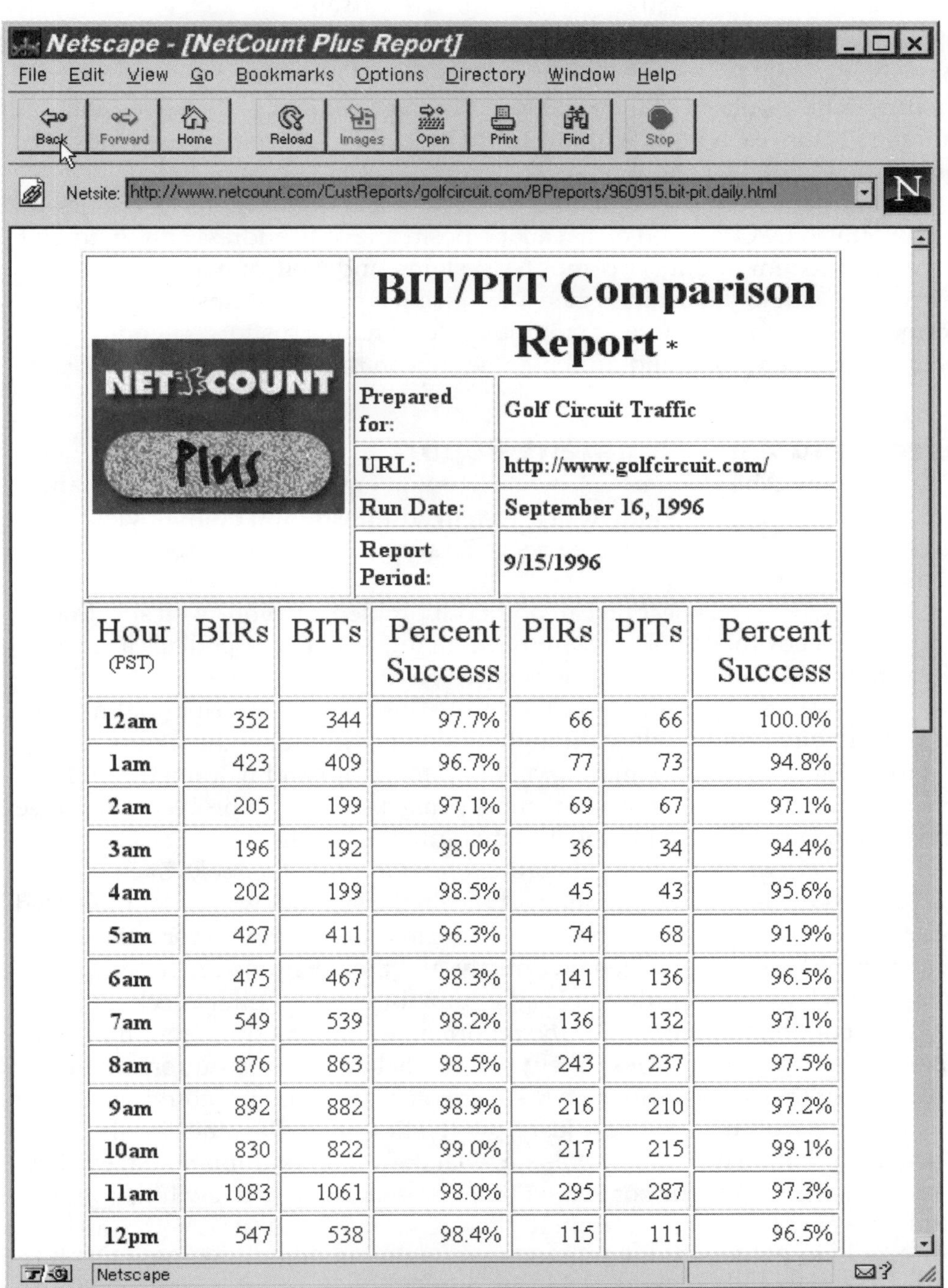

Hour (PST)	BIRs	BITs	Percent Success	PIRs	PITs	Percent Success
12am	352	344	97.7%	66	66	100.0%
1am	423	409	96.7%	77	73	94.8%
2am	205	199	97.1%	69	67	97.1%
3am	196	192	98.0%	36	34	94.4%
4am	202	199	98.5%	45	43	95.6%
5am	427	411	96.3%	74	68	91.9%
6am	475	467	98.3%	141	136	96.5%
7am	549	539	98.2%	136	132	97.1%
8am	876	863	98.5%	243	237	97.5%
9am	892	882	98.9%	216	210	97.2%
10am	830	822	99.0%	217	215	99.1%
11am	1083	1061	98.0%	295	287	97.3%
12pm	547	538	98.4%	115	111	96.5%

Figure 10-5. *The BIT/PIT Comparison Report*

shown in Figure 10-6. In addition to the number of unique visitors during each hour, NetCount also reports the number of those DPOs that are new unique visitors. At the bottom of the DPO Report is a graph that plots the number of unique visitors each hour, with the new DPOs represented by the shadowed area beneath total DPOs.

Again, NetCount's term distinct point of origin is analogous to visitors, but not exactly. A DPO is based on a reader's hostname or IP address only. It doesn't take into account multiuser computer systems, and NetCount employs their own algorithm to determine when a visit from a particular host has been terminated. For a more robust method of determining visitors, get onto NetCount's web pages and do some research on their HeadCount service.

Error and Abandonment Report

The Error and Abandonment Report (shown in Figure 10-7) is an important one, although perhaps more from a system administration point of view. This report shows the top five pages generating errors and the top five abandoned pages.

The top five errors section can point out problems in your web site. Maybe you have some broken links in your pages or image maps, or perhaps a graphic file is missing. Of the top five errors reported for the site in Figure 10-7, the top error-generating object is a Perl CGI script. This particular script (loglink.pl) records ad displays and click-throughs and logs them for archive purposes. On this particular day, we may have set up the script and worked through a few glitches; then again, the script simply may be having a problem. Whatever the reason, we will want to look into this situation further.

The second highest error-generating object in this table looks like pure garbage. There is no such silly-named file on the site, but we see 16 attempted accesses to this file. Of the remaining errors in the table, one is for a file with a directory name clearly spelled incorrectly, and the other two are for graphic images. Each of these entries looks like something we'd better check out.

The abandonment section of the report shows the top five pages that were aborted during loading. Most likely, people clicked the Stop button on their browsers or they clicked through the page on a link that they could see before it completed loading. We know just from the titles and filenames for these documents that each of these pages are what we call "click-through" pages (they have a very small amount of text on them presenting only a few link options). And we know that people routinely click on one of the options before all of the graphics have a chance to load. This doesn't bother us, but if a situation like this bothers you, you can always reduce the number and size of graphics on such a page to make it load more quickly. This way, fewer people will still be loading the page and graphics when they click on an option they're presented with.

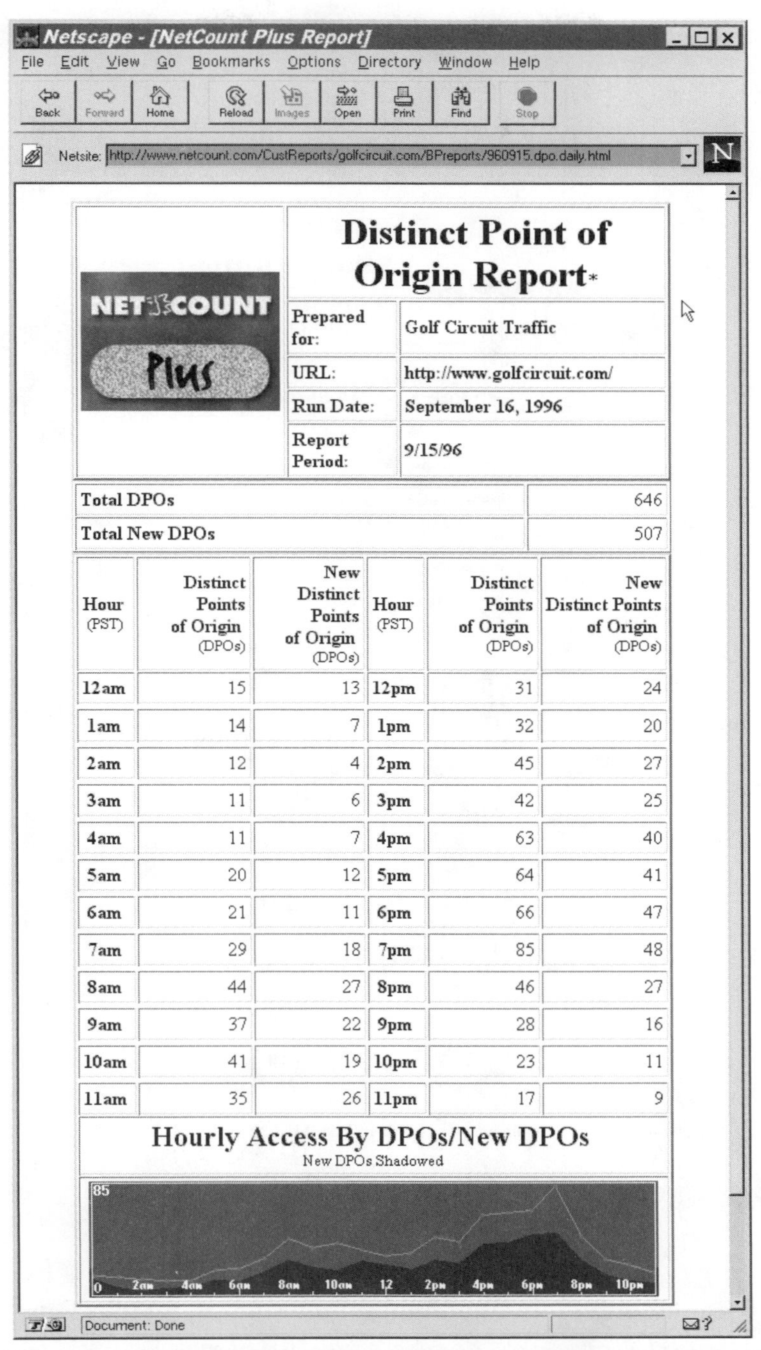

Distinct Point of Origin Report*

Prepared for:	Golf Circuit Traffic
URL:	http://www.golfcircuit.com/
Run Date:	September 16, 1996
Report Period:	9/15/96

Total DPOs	646
Total New DPOs	507

Hour (PST)	Distinct Points of Origin (DPOs)	New Distinct Points of Origin (DPOs)	Hour (PST)	Distinct Points of Origin (DPOs)	New Distinct Points of Origin (DPOs)
12am	15	13	12pm	31	24
1am	14	7	1pm	32	20
2am	12	4	2pm	45	27
3am	11	6	3pm	42	25
4am	11	7	4pm	63	40
5am	20	12	5pm	64	41
6am	21	11	6pm	66	47
7am	29	18	7pm	85	48
8am	44	27	8pm	46	27
9am	37	22	9pm	28	16
10am	41	19	10pm	23	11
11am	35	26	11pm	17	9

Hourly Access By DPOs/New DPOs

New DPOs Shadowed

Figure 10-6. *The Distinct Point of Origin (DPO) Report*

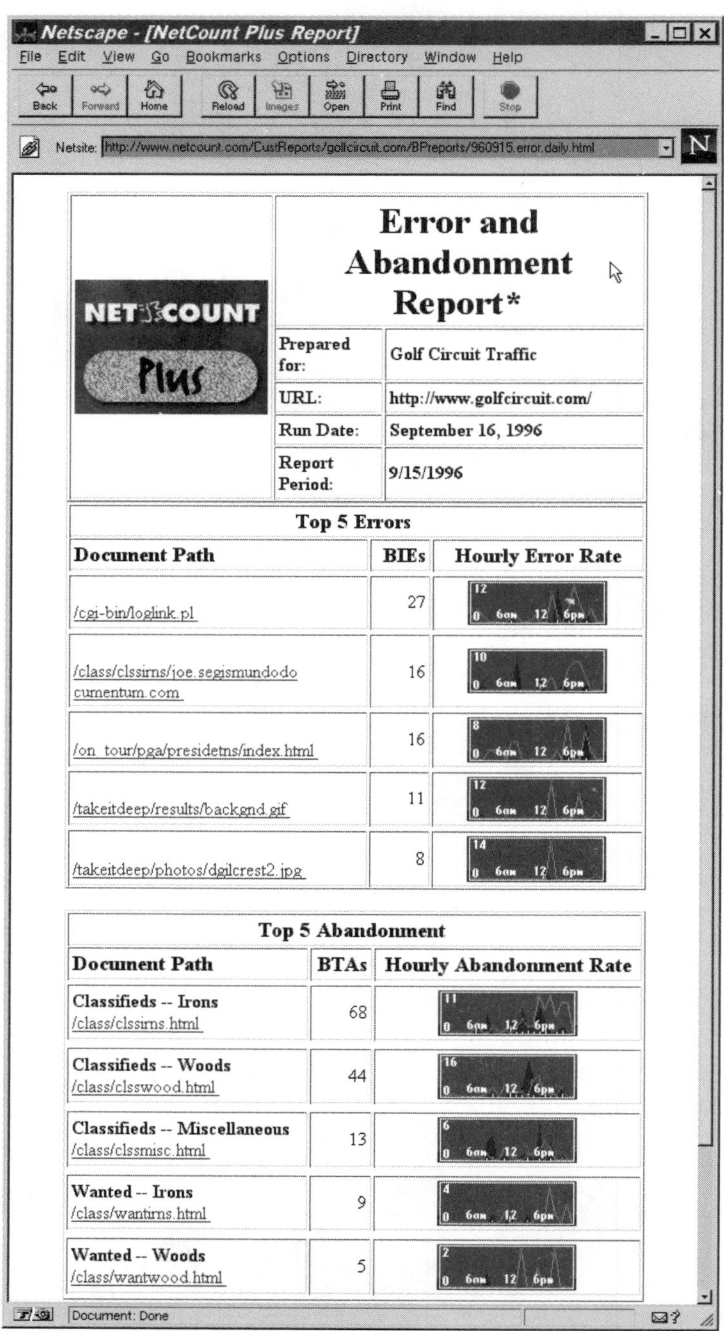

Figure 10-7. *NetCount's Error and Abandonment Report*

NetCount Weekly Reports

NetCount's weekly reports are accessible by clicking on any Sunday link in the Summary Reports calendar. Sundays are the only reports that include both daily reports and weekly reports (refer to Figure 10-2). The four weekly reports are listed immediately beneath the daily reports.

NetCount is contemplating monthly summary reports, but as yet, they haven't included them in either their Basic or Plus report sets.

Basic Report

The Basic Report summarizes the week's DPOs, PITs, and abandonments for the preceding week. Shown in Figure 10-8, the report summarizes these data points both numerically and graphically.

File Transfer By Type Report

The File Transfer By Type Report (see Figure 10-9) presents each type of object that you have on your site. NetCount gleans this information from a number of sources, including directory names, filename extensions, and the actual text of request headers. For each type of object, the table shows the total number of requests for each type of object, the number of those that were successful, and the success ratio.

Notice in this particular report that the HTML pages and images have had the highest success rates (96.82 percent and 99.72 percent, respectively), while CGI scripts had the lowest success ratio.

Root Domain Report

The Root Domain Report (see Figure 10-10) shows, for each day of the preceding week, the total number of DPOs and new DPOs by domain type. This shows you—to the extent possible—where your readers are coming from. The columns of the report represent the .com, .edu, .gov, .mil, .net, and .org root domains. The report also provides a column headed "INT" for international domains.

Of course, the dubious accuracy of this report is evidenced by the fact that one of the largest categories is always the Unknown column. This is because so many web users are using computers for which they've never registered a domain name or they have an Internet Service Provider (ISP) that assigns IP addresses dynamically from a pool.

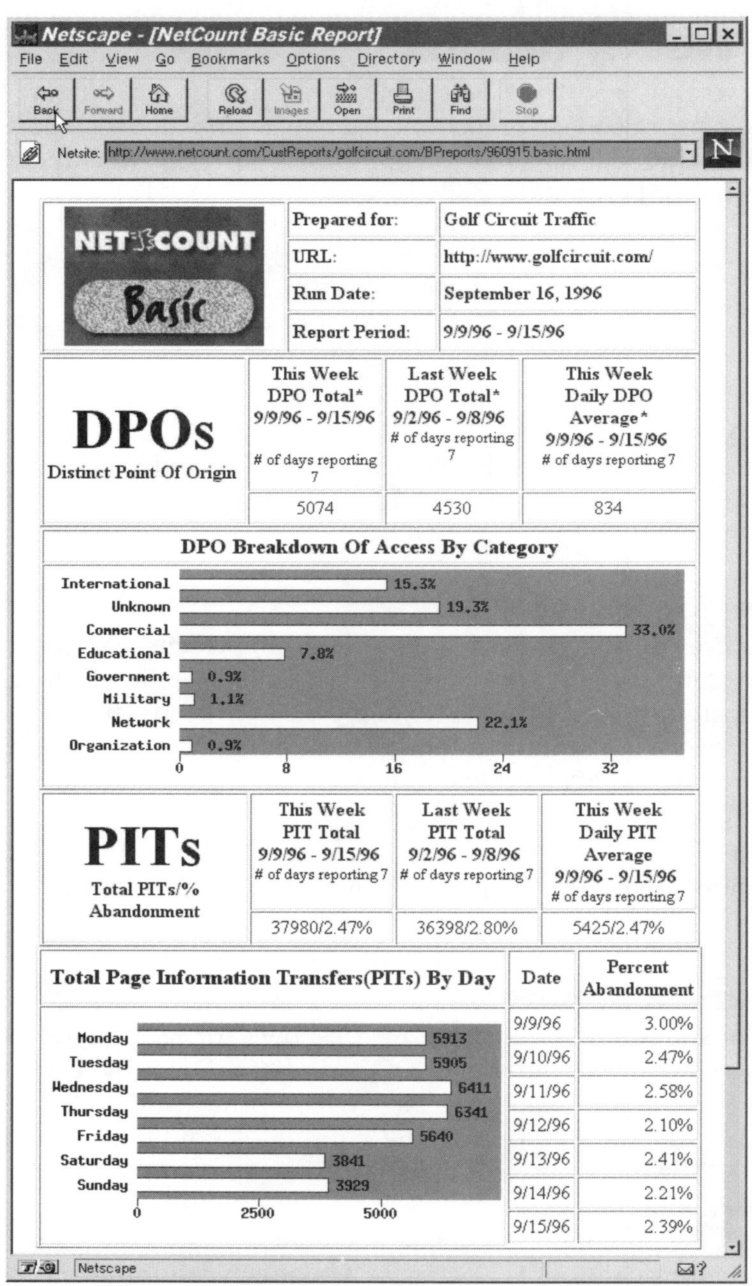

Figure 10-8. *The Basic Report*

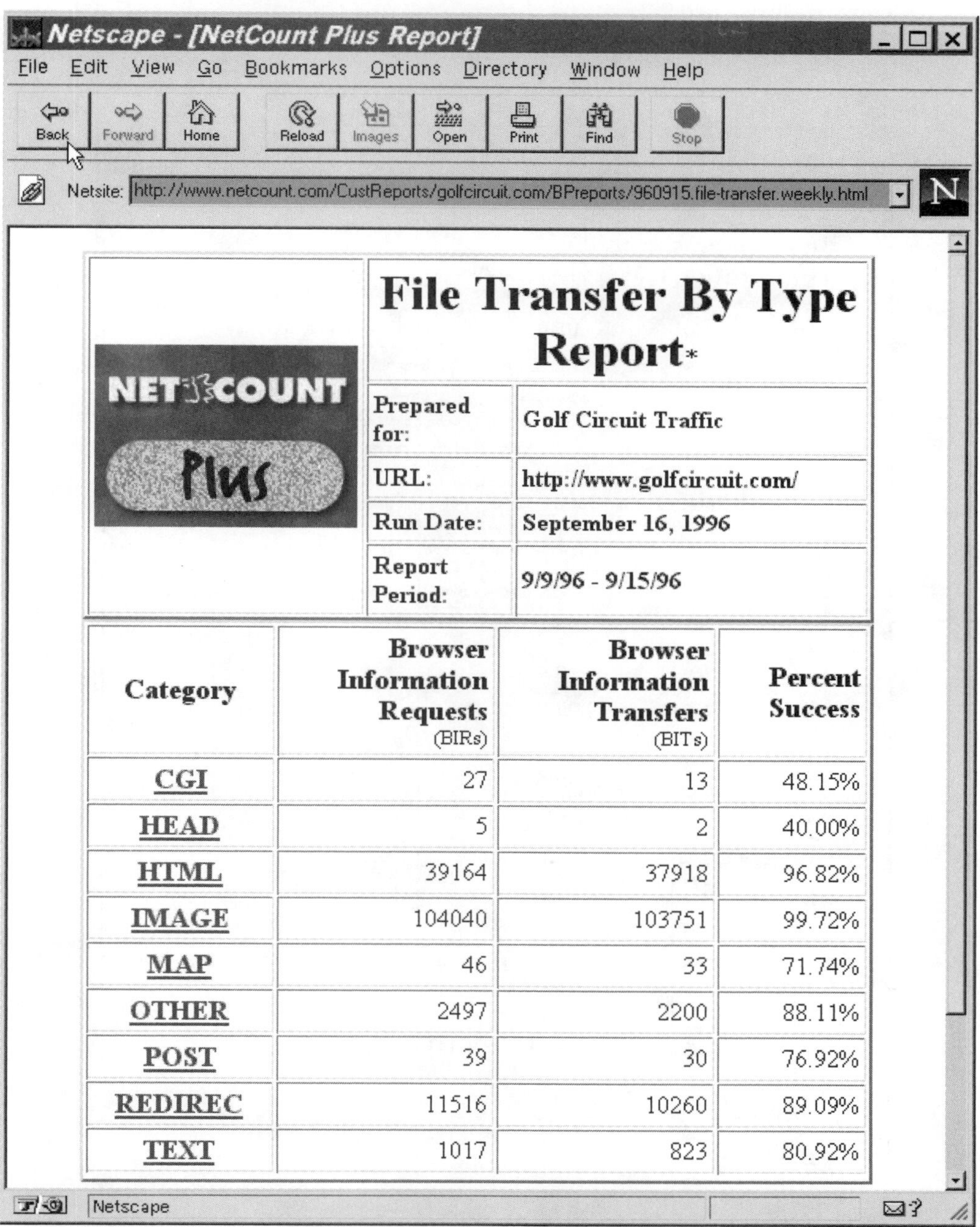

File Transfer By Type Report*

Prepared for:	Golf Circuit Traffic
URL:	http://www.golfcircuit.com/
Run Date:	September 16, 1996
Report Period:	9/9/96 - 9/15/96

Category	Browser Information Requests (BIRs)	Browser Information Transfers (BITs)	Percent Success
CGI	27	13	48.15%
HEAD	5	2	40.00%
HTML	39164	37918	96.82%
IMAGE	104040	103751	99.72%
MAP	46	33	71.74%
OTHER	2497	2200	88.11%
POST	39	30	76.92%
REDIREC	11516	10260	89.09%
TEXT	1017	823	80.92%

Figure 10-9. *The File Transfer By Type Report*

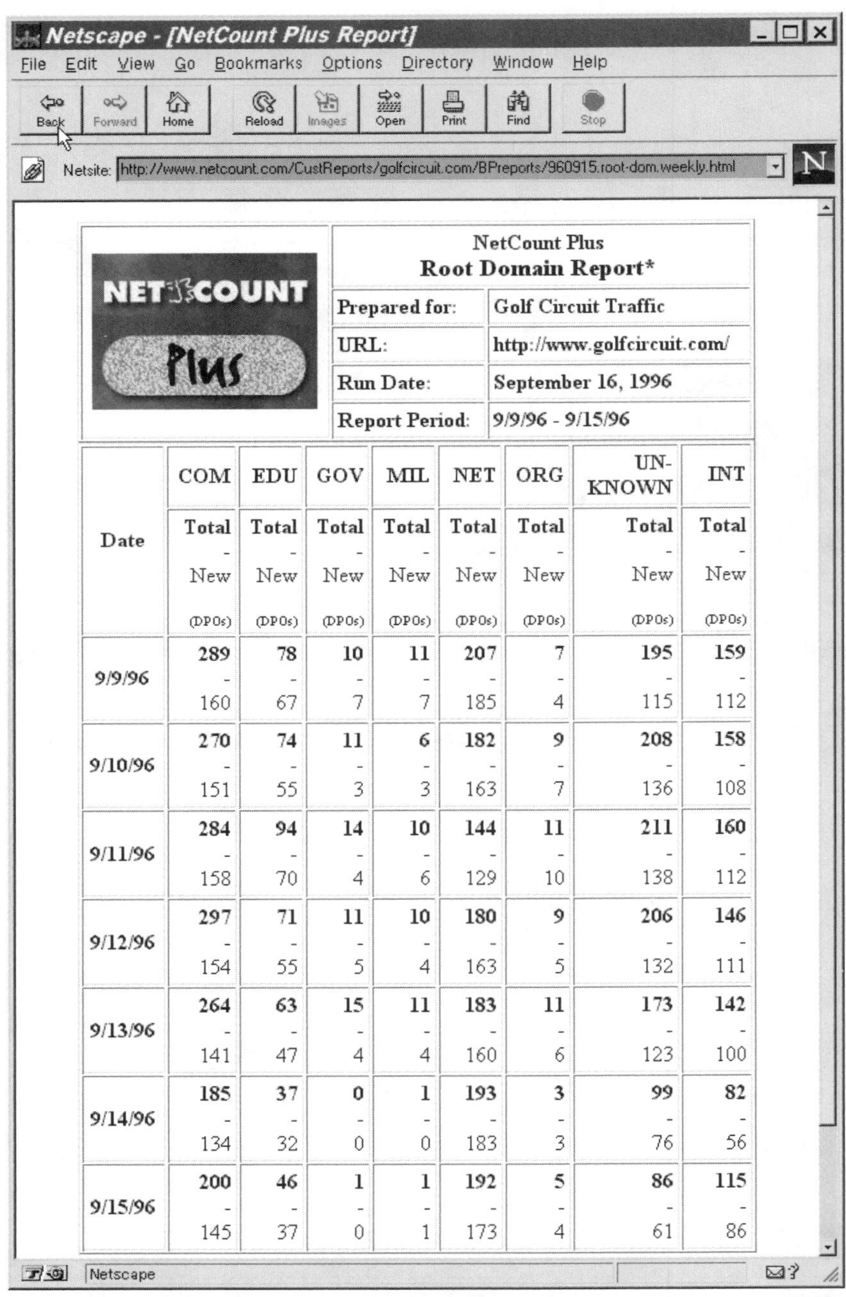

NetCount Plus
Root Domain Report*

Prepared for:	Golf Circuit Traffic
URL:	http://www.golfcircuit.com/
Run Date:	September 16, 1996
Report Period:	9/9/96 - 9/15/96

Date	COM	EDU	GOV	MIL	NET	ORG	UN-KNOWN	INT
	Total - New (DPOs)	**Total** - New (DPOs)	**Total** - New (DPOs)	**Total** - New (DPOs)	**Total** - New (DPOs)	**Total** - New (DPOs)	**Total** - New (DPOs)	**Total** - New (DPOs)
9/9/96	**289** - 160	**78** - 67	**10** - 7	**11** - 7	**207** - 185	**7** - 4	**195** - 115	**159** - 112
9/10/96	**270** - 151	**74** - 55	**11** - 3	**6** - 3	**182** - 163	**9** - 7	**208** - 136	**158** - 108
9/11/96	**284** - 158	**94** - 70	**14** - 4	**10** - 6	**144** - 129	**11** - 10	**211** - 138	**160** - 112
9/12/96	**297** - 154	**71** - 55	**11** - 5	**10** - 4	**180** - 163	**9** - 5	**206** - 132	**146** - 111
9/13/96	**264** - 141	**63** - 47	**15** - 4	**11** - 4	**183** - 160	**11** - 6	**173** - 123	**142** - 100
9/14/96	**185** - 134	**37** - 32	**0** - 0	**1** - 0	**193** - 183	**3** - 3	**99** - 76	**82** - 56
9/15/96	**200** - 145	**46** - 37	**1** - 0	**1** - 1	**192** - 173	**5** - 4	**86** - 61	**115** - 86

Figure 10-10. *The Root Domain Report*

Top Ten Organizations By Sub Domain Report

The Top Ten Organizations By Sub Domain Report shows, for each root domain, the top ten subdomains accessing your site. For example, in Figure 10-11, you can see the top ten commercial domains (in the .com root domain), followed by the top ten educational domains (in the .edu domain), and so on.

Although NetCount doesn't offer the breadth of reports and statistics that some log analysis software packages do, they do offer the basics—the most important reports and statistics for analyzing your traffic flow and tracking down errors. Without question, though, NetCount offers the most wholesome comparisons between requests and actual successful transfers. For a better idea of how NetCount stacks up against commercial and freeware/shareware log analysis software, see Appendix A.

I/PRO

NetCount's main competition in the log analysis service industry is Internet Profiles Corporation (I/PRO) of San Francisco. I/PRO has teamed up with Nielsen Media Research (the TV ratings company) to offer three services to web site owners and administrators:

- I/COUNT
- I/AUDIT
- I/CODE

The I/COUNT service is similar in many ways to NetCount's tracking service. I/PRO provides a small amount of software that runs on your server and sends your transfer log data back to headquarters. Unlike NetCount, though, I/PRO's software doesn't send your data in hourly; I/PRO does only a single transfer each day.

I/PRO's pricing for I/COUNT is also considerably different than that of NetCount. I/PRO charges $200 per month for sites with fewer than (only) 5,000 hits per day. For sites with between 100,000 and 300,000 hits per day, you should expect to pay $3,000 monthly.

Unfortunately, I/PRO chose not to provide an evaluation account for this book, so we don't have any detailed information about their service or sample reports to show you. I/PRO did, however, provide two interesting multimedia presentations on the CD-ROM at the back of the book. There's not much detail there either, but it's a starting place for you.

I/PRO's I/AUDIT service takes log analysis up a notch. With this service, I/PRO will run additional software and algorithms against your log data to

Netscape - [NetCount Plus Report]

File Edit View Go Bookmarks Options Directory Window Help

Back Forward Home Reload Images Open Print Find Stop

Netsite: http://www.netcount.com/CustReports/golfcircuit.com/BPreports/960915.top-10.weekly.html

NET COUNT Plus

Top Ten Organizations By Sub Domain Report*

Prepared for:	Golf Circuit Traffic
URL:	http://www.golfcircuit.com/
Run Date:	September 16, 1996
Report Period:	9/9/96 - 9/15/96

Commercial

#	Domain	Company	BITs	#	Domain	Company	BITs
1.	**compuserve.com**	CompuServe, Inc.	4065	6.	**erols.com**	Erols	820
2.	**netcom.com**	NETCOM	3917	7.	**mindspring.com**	MindSpring Enterprises, Inc.	757
3.	**aol.com**	America Online, Inc.	3255	8.	**prodigy.com**	Prodigy Services Company	705
4.	**znet.com**	zNET	993	9.	**boeing.com**	The Boeing Company	700
5.	**telia.com**		867	10.	**infoseek.com**	Infoseek	615

Education

#	Domain	School	BITs	#	Domain	School	BITs
1.	**tamu.edu**	Texas A&M University	579	6.	**berkeley.edu**	University of California at Berkeley	364
2.	**purdue.edu**	Purdue University	489	7.	**psu.edu**	Pennsylvania State University	350
3.	**ohio-state.edu**	Ohio State University	449	8.	**okstate.edu**	Oklahoma State University	285
4.	**stanford.edu**	Stanford University	365	9.	**unc.edu**	University of North Carolina at Chapel Hill	244
5.	**uiuc.edu**	University of Illinois at Urbana-Champaign	364	10.	**gatech.edu**		208

Netscape

Figure 10-11. *The Top Ten Organizations By Sub Domain Report*

NIELSEN MEDIA RESEARCH
299 Park Avenue
New York, New York 10171-0074
(212) 708-7500, FAX: (212) 708-7600
URL: http://www.nielsenmedia.com

INTERNET PROFILES CORPORATION
785 Market Street, 13th Floor
San Francisco, CA 94103
(415) 975-5800, FAX: (415) 975-5818
URL: http://www.ipro.com

I/AUDIT™ STATEMENT

The purpose of this I/AUDIT report is to verify usage of a Web site and to state usage in audited terms. This information is to be used by Content Publishers, Advertisers, Media Buyers/Planners, and Web Developers.

Completed November 9, 1995

Summary of Usage and Activity for **October 1995**:

individual

Individual, Inc.
8 New England Executive Park West
Burlington, MA 01803
(617) 273-6000
URL:http://www.newspage.com

SUMMARY OF USAGE	October	September
Total Visits* per Audit Period	214,789	163,381
Average Visits per Day	6,929	5,446
Total Hits per Audit Period	6,326,595	5,511,400
Total Qualified Hits	6,026,970	5,244,535
Average Qualified Hits per Day	194,418	174,818
Average Qualified Hits per Visit	28.1	32.1
Average Visit Length (minutes)	6:00	6:44

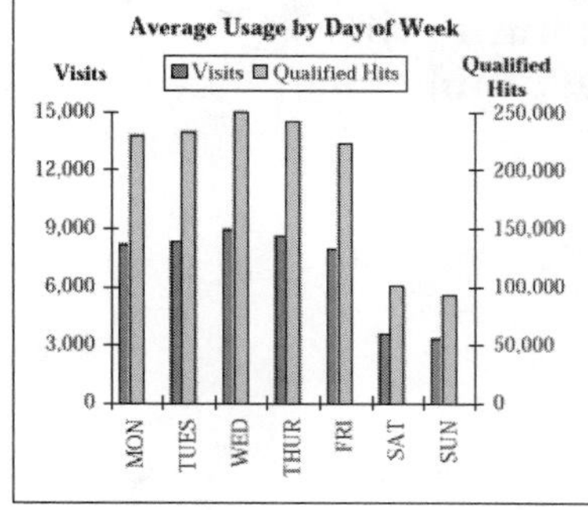

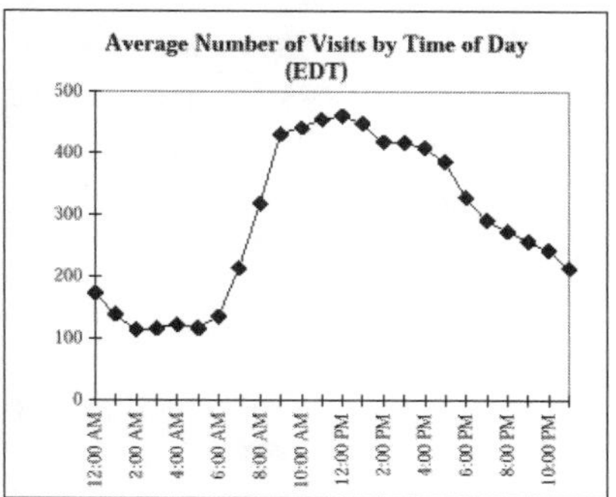

*Refer to last page for definitions and explanations

Return to Customer I/AUDITs

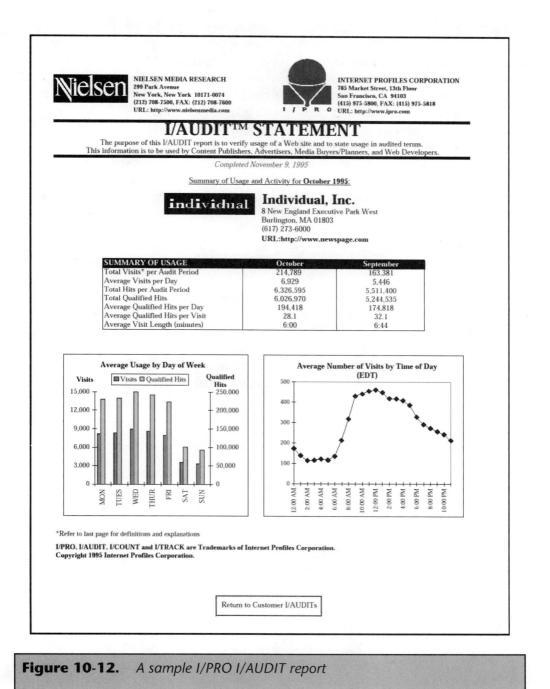

Figure 10-12. *A sample I/PRO I/AUDIT report*

reveal problem areas, such as days when your server might have been down or your Internet connection was broken. They will review each of these anomalies and even call you to ask about them, if necessary. Once I/PRO has reviewed the entire month's statistics and is confident that there are no unexplained unusual fluctuations, they produce a printed "audit report" (shown on the previous page in Figure 10-12), which sports the Nielsen and I/PRO logos.

The I/CODE service is a universal user registration system. It has evolved from the days when so many site administrators tried to collect their own user demographic information through their own sites with user registration. The downside to doing this was enormous. As the volume of fun places to visit on the web increased exponentially, the web user's patience for registering at individual web sites went into the tank. I/PRO's idea with I/COUNT is to provide a single, centralized location for users of many sites to register. Once they've done this, they can simply enter their I/CODE account number at sites that use I/CODE for access to special parts of the site or to be eligible for special promotions.

Though you may loose a little in terms of the flexibility of your reporting, there's certainly a great deal to be said for the simplicity of letting others analyze your traffic for you. This simplicity does come at a higher price. But if that fits your budget, you may find it well worth the price—especially if independently generated third-party statistics are a high priority to you.

In the next chapter, we will look at several popular shareware and freeware log analysis packages and how they stack up against both commercial software and log analysis services.

Chapter Eleven

Free and Shareware Statistics Software

Not long ago, just about every web server log analysis tool you could have found would have been freeware or shareware. There were plenty of programs to choose from, but there just wasn't a commercial market servicing this niche. However, since companies like e.g. Software, Intersé, net.Genesis, I/PRO, and others have come around, this situation is beginning to change.

The professional developers have been good for log analysis. As they compete against each other to produce the best products and services, the general level of quality has increased and prices have fallen. But there is a cost (besides the obvious cost of commercial packages). Freeware and shareware developers seem to have become discouraged, and many of them have put their projects on the back burner as they realize that they can't possibly compete with the development efforts of whole companies with teams of programmers. Perhaps it's for this reason that free log analysis tools are becoming stale.

Nevertheless, there are still many freeware and shareware programs out there, and many site administrators prefer using them to the new commercial packages. There are dozens of potentially viable free products that you can use to run server statistics, but most system administrators have settled on a few of them that stand out above the rest.

In this chapter, we will look at several of the most popular free and shareware packages, and we will discuss others in more general terms.

wwwstat

wwwstat is an old favorite with Unix system administrators who have been around for a while. wwwstat is a Perl program that was written by Roy Fielding of the University of California, Irvine. As Perl programs go, wwwstat is fairly simple. It's 1,100 lines of Perl code, and it uses no external libraries and only one external data file—a file with the root domain country codes for the countries of the world.

An obvious improvement that could be made to wwwstat is to add another external file: a configuration file that holds the values the program needs to process log files. As it is now, you have to edit the Perl program itself to put in values for the following:

- The title of the report
- The path to the country codes file

- The path and filename for the server log file
- The domain name of the site

Besides it being somewhat cumbersome to have to edit the program code itself, there's an even more onerous drawback to wwwstat. The program can only process the single log file that you tailor it to process. If you virtually host a dozen web sites, for example, you will have to actually make 12 different copies of the program—each tailored to its own site and log file.

If you are a Perl programmer, it wouldn't be a big job to customize wwwstat to function generically and then take its information from a configuration file. But you would have to find the time to do it.

wwwstat does have some strengths. For one, it has a host of command-line options that makes it convenient to run it from a command line or out of a Unix system's crontab file. You simply execute the program and redirect its output to the filename you want for your reports.

The main strength of wwwstat is the detail of its report. Figure 11-1 shows the top of the report created by wwwstat for the golf site. At the top, it offers links down to each section of the report. This may seem like a minor feature, but you will soon find that the report includes so much detail that you will want all the aids you can use to help you get around. In fact, the report from wwwstat is so detailed that for the golf site (for the same month we ran the commercial analysis software), wwwstat produced a report of over one-half megabyte of pure HTML.

Some site administrators like this level of detail; they feel that it's important to be able to see the actual number of requests for every single file in the web site. This isn't a big issue when you've got a couple hundred pages in your web site, but when you have a couple thousand pages, it can become a liability. Other administrators want summary reports; they feel that producing summary reports is the function of log analysis software. So, apparently, the beauty of detail is in the eye of the beholder.

wwwstat produces six sections in its report:

- Summary statistics
- Daily transmission statistics
- Hourly transmission statistics
- Total transfers by client domain
- Total transfers by reversed subdomain
- Total transfers from each archive section

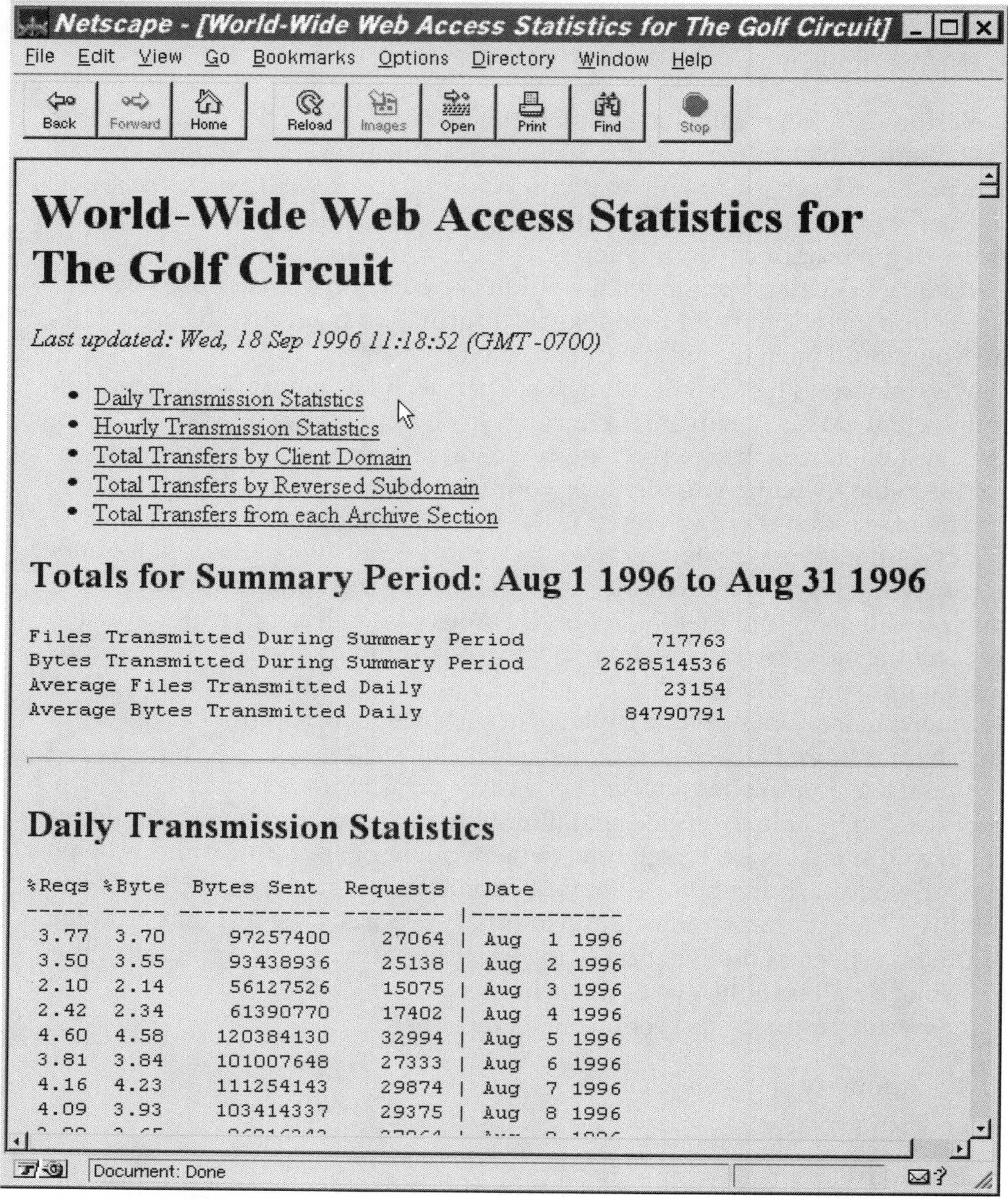

Figure 11-1. *wwwstat produces a no-frills summary of your traffic with plenty of details*

The summary statistics section is at the very top of the report and is visible in Figure 11-1. It reports on the following:

- Files transmitted during the summary period
- Bytes transmitted during the summary period
- Average files transmitted daily
- Average bytes transmitted daily

Notice that there is no distinction at all between hits and visits. wwwstat makes no attempt to count visits or points of origin. Therefore, it has no way to report on vital statistics like the number of visitors during a period, average number of page views during a visit, average view duration, and so on.

Neither does wwwstat make any distinction between requests and successful transfers. It reports only files and bytes transferred during the period and the resulting daily averages. Another improvement wwwstat begs for is the addition of commas—without them, it can be pretty hard to read numbers with eight and nine digits.

As mentioned earlier, the main strength of wwwstat (besides its simplicity) is its detail. That is, if detail is what you want. The Daily Transmission Statistics section (shown in Figure 11-2) reports, for each day of the report period, the total number of requests and bytes transferred.

The Hourly Transmission Statistics section (see Figure 11-3) reports the average number of requests and bytes transferred for each hour of the day throughout the period. Both the Daily Transmission Statistics section and the Hourly Transmission Statistics section report the percentages for transfers and bytes of each day (or hour) for the whole report period.

The three remaining sections of the wwwstat report are for total transfers—by client domain, by reversed subdomain, and by archive section. The Total Transfers by Client Domain section shows the number of requests and bytes sent for each root domain. These will be mostly country codes (from the country codes file) if you attract international attention; if not, they will be the usual .com, .edu, .mil, and so on.

The Total Transfers by Reversed Subdomain section reports the same data (number of requests and bytes transferred) for each domain that accessed your server during the report period. It's not immediately clear why the domain names are reversed—unless it's just to make the report sort the domains by root domain type.

The Total Transfers from each Archive Section reports (again) the number of raw requests and bytes sent for each object in your web site that was accessed

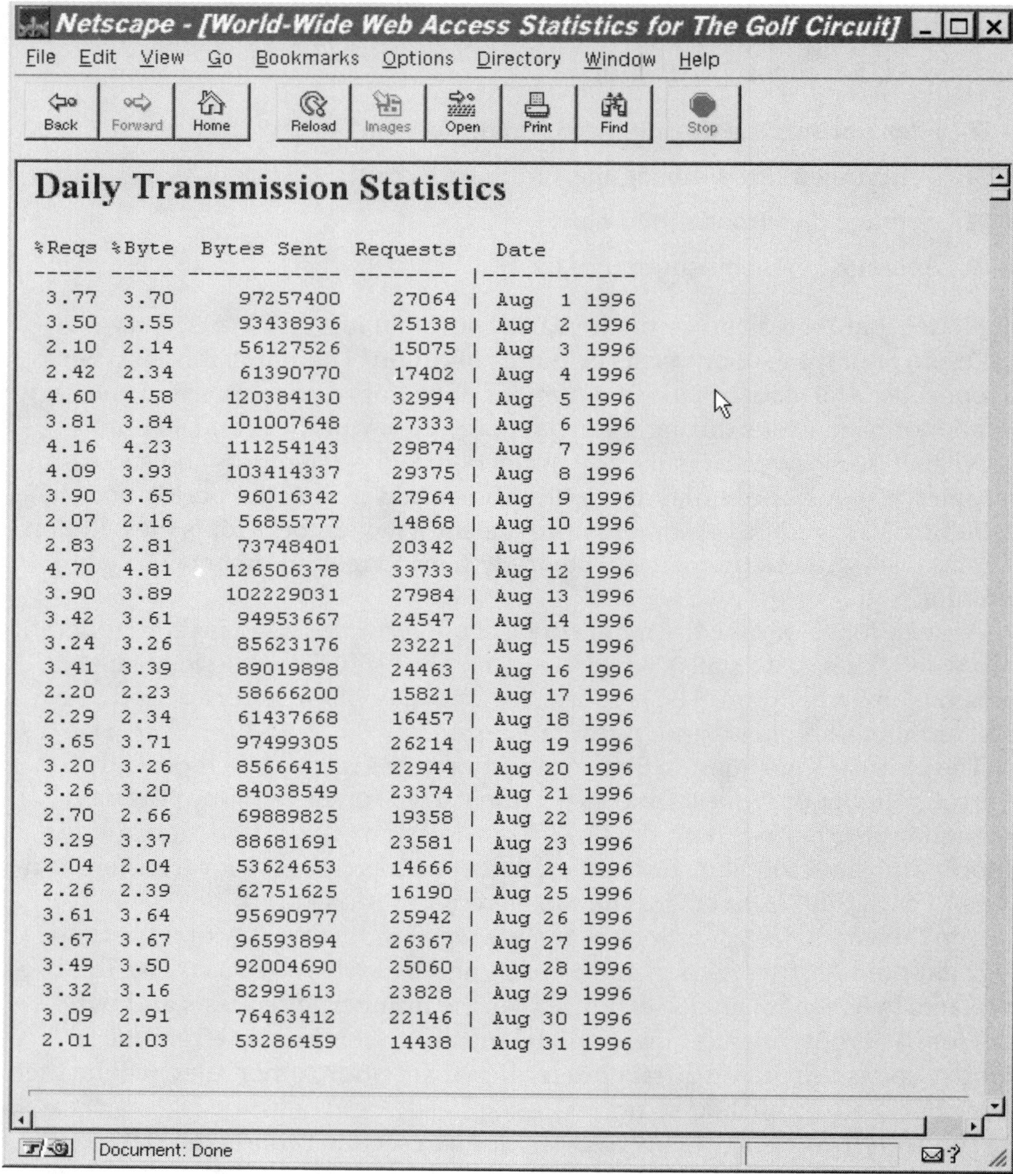

Daily Transmission Statistics

%Reqs	%Byte	Bytes Sent	Requests	Date
3.77	3.70	97257400	27064	Aug 1 1996
3.50	3.55	93438936	25138	Aug 2 1996
2.10	2.14	56127526	15075	Aug 3 1996
2.42	2.34	61390770	17402	Aug 4 1996
4.60	4.58	120384130	32994	Aug 5 1996
3.81	3.84	101007648	27333	Aug 6 1996
4.16	4.23	111254143	29874	Aug 7 1996
4.09	3.93	103414337	29375	Aug 8 1996
3.90	3.65	96016342	27964	Aug 9 1996
2.07	2.16	56855777	14868	Aug 10 1996
2.83	2.81	73748401	20342	Aug 11 1996
4.70	4.81	126506378	33733	Aug 12 1996
3.90	3.89	102229031	27984	Aug 13 1996
3.42	3.61	94953667	24547	Aug 14 1996
3.24	3.26	85623176	23221	Aug 15 1996
3.41	3.39	89019898	24463	Aug 16 1996
2.20	2.23	58666200	15821	Aug 17 1996
2.29	2.34	61437668	16457	Aug 18 1996
3.65	3.71	97499305	26214	Aug 19 1996
3.20	3.26	85666415	22944	Aug 20 1996
3.26	3.20	84038549	23374	Aug 21 1996
2.70	2.66	69889825	19358	Aug 22 1996
3.29	3.37	88681691	23581	Aug 23 1996
2.04	2.04	53624653	14666	Aug 24 1996
2.26	2.39	62751625	16190	Aug 25 1996
3.61	3.64	95690977	25942	Aug 26 1996
3.67	3.67	96593894	26367	Aug 27 1996
3.49	3.50	92004690	25060	Aug 28 1996
3.32	3.16	82991613	23828	Aug 29 1996
3.09	2.91	76463412	22146	Aug 30 1996
2.01	2.03	53286459	14438	Aug 31 1996

Figure 11-2. *The wwwstat Daily Transmission Statistics section shows the number of requests and bytes transferred for each day of the report period and the percentage for each*

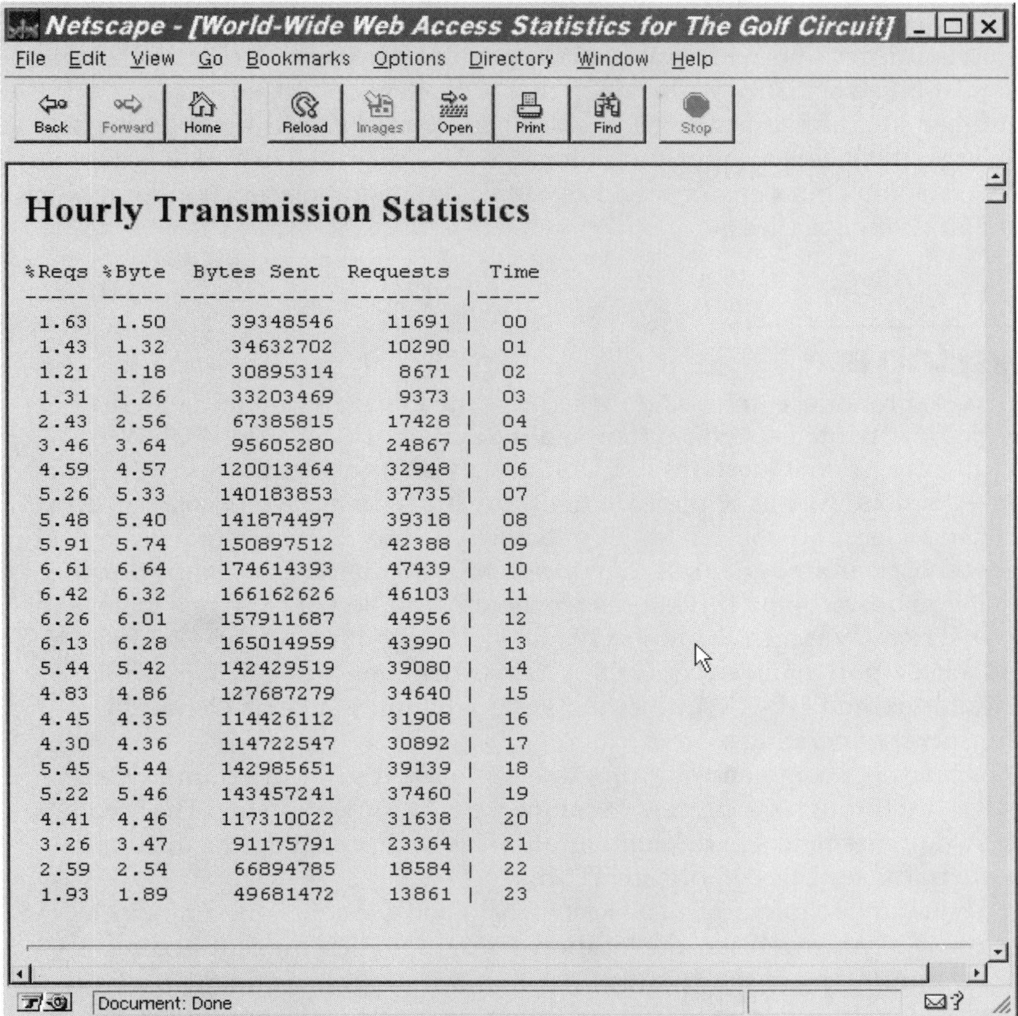

Netscape - [World-Wide Web Access Statistics for The Golf Circuit]

Hourly Transmission Statistics

%Reqs	%Byte	Bytes Sent	Requests	Time
1.63	1.50	39348546	11691	00
1.43	1.32	34632702	10290	01
1.21	1.18	30895314	8671	02
1.31	1.26	33203469	9373	03
2.43	2.56	67385815	17428	04
3.46	3.64	95605280	24867	05
4.59	4.57	120013464	32948	06
5.26	5.33	140183853	37735	07
5.48	5.40	141874497	39318	08
5.91	5.74	150897512	42388	09
6.61	6.64	174614393	47439	10
6.42	6.32	166162626	46103	11
6.26	6.01	157911687	44956	12
6.13	6.28	165014959	43990	13
5.44	5.42	142429519	39080	14
4.83	4.86	127687279	34640	15
4.45	4.35	114426112	31908	16
4.30	4.36	114722547	30892	17
5.45	5.44	142985651	39139	18
5.22	5.46	143457241	37460	19
4.41	4.46	117310022	31638	20
3.26	3.47	91175791	23364	21
2.59	2.54	66894785	18584	22
1.93	1.89	49681472	13861	23

Document: Done

Figure 11-3. *The Hourly Transmission Statistics show the average number of requests and bytes transferred for each hour of the day and the percentage for each*

during the period. Here's where you look to see how many people viewed a particular page or graphic (such as an ad banner). This section would be more valuable if it were sorted in descending order of the number of requests, but, alas, it's sorted alphabetically by the filename of each object.

In summary, wwwstat is an adequate log analysis program if all you want is cursory information about how people use your web site. It offers no visit-based statistics at all, no top five or top ten reports, no graphics, and a tremendous amount of detail. It's also hard to read if your numbers are at all large. But, being a command-line program and written in Perl, it is easy to integrate into your daily or weekly system administration functions of rotating and archiving log files.

wusage

wusage (pronounce dub-usage) is a great little log analysis package written by Thomas Boutell, a book author and the author of the World Wide Web Frequently Asked Questions list. Unlike wwwstat, wusage comes with a price tag—it's shareware as opposed to freeware. But a license won't cost you an arm and a leg, and you can try it out before you buy it. A copy of wusage will set you back around $75 (less for educational and nonprofit organizations).

If you use a computer, chances are there's a version of wusage available for your system. wusage runs under Windows 95, Windows NT, OS/2, MS-DOS, and a host (pun intended) of Unix systems, including Sun, BSDI, FreeBSD, IRIX, Linux, and HP-UX. Interestingly, the program functions identically under every operating system.

wusage is a command-line program. Actually, it's two programs. One walks you through the process of setting up a configuration file. The second uses your parameters in the configuration file to process your log files and create traffic statistics reports in HTML.

The setup program (called makeconf) asks you a series of questions, including the name of your web site, the location of your log files, the output directory for its reports, and so on. Figure 11-4 shows this program running in a DOS window under Windows 95. Since there are separate versions of wusage for DOS, Windows 95 and NT, and OS/2, one would presume that there are some actual differences in the program. Whatever these differences are, however, they aren't apparent to the naked eye. The program looks and feels under DOS exactly like it does under Unix.

The makeconf program sets up most of the parameters that the program needs to run. There are a few options you might want to tweak to customize your reports. For example, by default, wusage runs statistics on all objects—

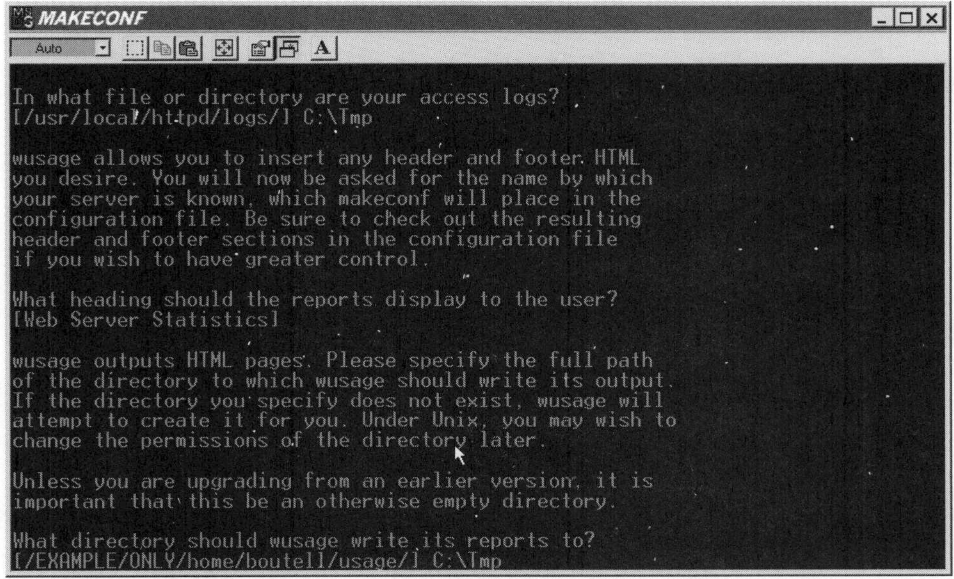

MAKECONF
Auto
In what file or directory are your access logs?
[/usr/local/httpd/logs/] C:\Tmp
wusage allows you to insert any header and footer HTML
you desire. You will now be asked for the name by which
your server is known, which makeconf will place in the
configuration file. Be sure to check out the resulting
header and footer sections in the configuration file
if you wish to have greater control.
What heading should the reports display to the user?
[Web Server Statistics]
wusage outputs HTML pages. Please specify the full path
of the directory to which wusage should write its output.
If the directory you specify does not exist, wusage will
attempt to create it for you. Under Unix, you may wish to
change the permissions of the directory later.
Unless you are upgrading from an earlier version, it is
important that this be an otherwise empty directory.
What directory should wusage write its reports to?
[/EXAMPLE/ONLY/home/boutell/usage/] C:\Tmp

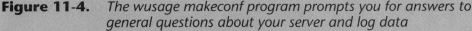

Figure 11-4. *The wusage makeconf program prompts you for answers to general questions about your server and log data*

pages, graphics, Java applets, and so on. When you generate statistics with this default, you may find that your top ten documents list includes graphics. If you use a lot of graphics on many pages (like a background image or an imagemap graphic), you may find these graphics files in the list. If you want to see only HTML pages, you can tell wusage to ignore graphics files (or any kind of file) by editing the configuration file (named wusage.conf by default) and specifying filename extensions to ignore.

The standard summary report (shown in Figure 11-5) has some general information at the top followed by several tables, most of which are accompanied by a pie chart. The tabular information in each table ranks the top ten candidates (or another number, which you can specify in the configuration file) for each table. The columnar data included in each table is the number of accesses and the number of bytes transferred. (In the configuration file you can specify whether the top ten candidates for each table are based on the number of requests or the volume of bytes transferred.)

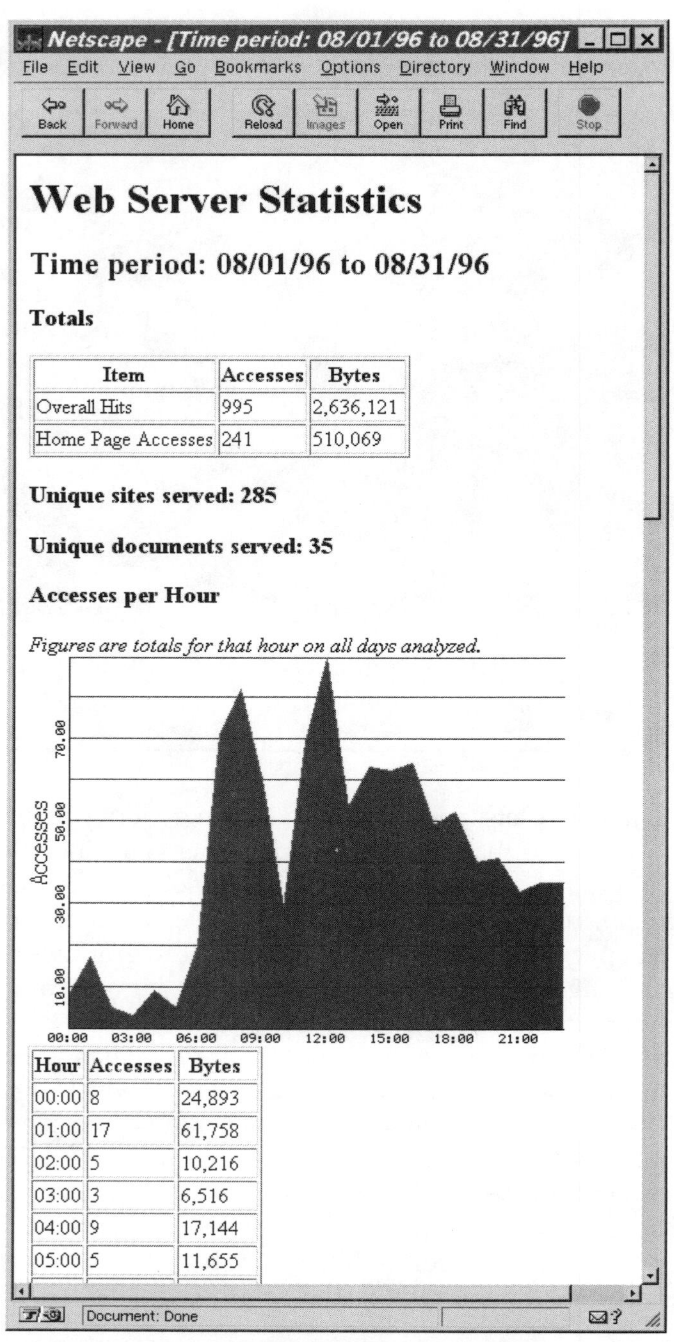

Figure 11-5. *The wusage summary report*

The wusage tables report on the following:

- Accesses (or bytes transferred) by the hour of the day
- Top ten documents by access count (or bytes transferred)
- Top ten sites by access count (or bytes transferred)
- Top ten domains by access count (or bytes transferred)
- Number of hits generating each server result code

The general information at the top of the summary report (which you can see in Figure 11-5) includes the total number of requests and bytes transferred, the number of requests for the site home page, the total number of unique sites served, and the number of unique documents served.

With the exception of reporting on the number of unique sites served, wusage is completely hit-based. That is, the statistics it generates include none of the extended statistics that you can compute when you know the exact number of visitors to a site (such as average number of page views per visit, average duration of page views, and so on). Although the number of unique sites served during the time period doesn't exactly equate to visits, it isn't a bad estimation of visits. You could call this number visits if you don't mind ignoring the effects of multiuser computer systems and repeat visits. After all, even some of the biggies—like NetCount—don't do more than this in their standard reports. (Recall that NetCount calls these DPOs, for distinct point of origin.)

One innovative addition is the Accesses by Result Code table in the Top 10 Domains by Access Count report. Shown in Figure 11-6, this table shows the number of requests that resulted in each access code. You can see at a glance the relative number of status 200 requests (successful transfers), cached documents (code 304—Not Modified), and error codes. Usually, the code 404 (Not found) is the most common error code.

You can configure wusage to give you additional information on error conditions. By including the keyword "notfound" in the configuration file, you can make wusage include a table with the top ten (or however many you want) error requests. This is a valuable tool for tracking down stale links in your site, notifying you of changes you should make with search engines, or redirecting directives you should put in your server configuration file.

In summary, wusage is a simple package with informative and clean reports. Its command-line operation makes it ideal for integrating with system administrative functions and automating log processing—even for multiple sites. The biggest drawback to wusage is its lack of visit focus and lack of details. If you like knowing exactly how many requests were made for every page in your site, this package won't do. For quick, easy, clean, and fully automated summary reports, wusage shines.

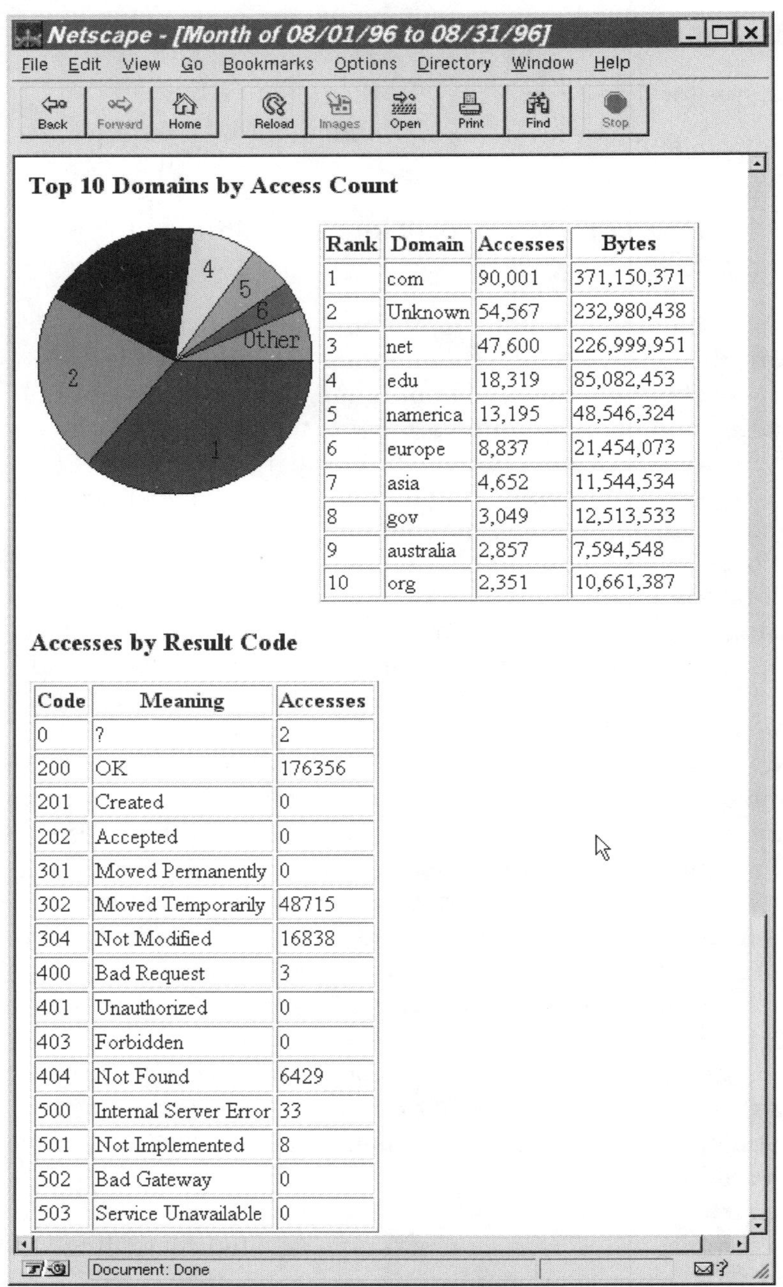

Top 10 Domains by Access Count

Rank	Domain	Accesses	Bytes
1	com	90,001	371,150,371
2	Unknown	54,567	232,980,438
3	net	47,600	226,999,951
4	edu	18,319	85,082,453
5	namerica	13,195	48,546,324
6	europe	8,837	21,454,073
7	asia	4,652	11,544,534
8	gov	3,049	12,513,533
9	australia	2,857	7,594,548
10	org	2,351	10,661,387

Accesses by Result Code

Code	Meaning	Accesses
0	?	2
200	OK	176356
201	Created	0
202	Accepted	0
301	Moved Permanently	0
302	Moved Temporarily	48715
304	Not Modified	16838
400	Bad Request	3
401	Unauthorized	0
403	Forbidden	0
404	Not Found	6429
500	Internal Server Error	33
501	Not Implemented	8
502	Bad Gateway	0
503	Service Unavailable	0

Figure 11-6. *The Accesses by Result Code table*

VBStats

VBStats is a public domain program originally written by Bob Denny. Interestingly, he credits Tom Boutell's wusage program as his inspiration for writing VBStats, although the program falls well short of the flexibility and usefulness of wusage. But there are some nice things to be said about VBStats. For one, being in the public domain, it's completely free. And if you like to tinker with Visual Basic, you can get the source code and customize it to fit your needs.

VBStats is a 16-bit Windows program, although it really doesn't matter much that it's a Windows program, because for the most part there is no user interface. The program includes five executable programs:

- A setup wizard
- A log file import program
- A query restriction module
- The report program
- A database maintenance utility

One of the first limitations of VBStats that you will run into is that an installation of the program can process and produce output for only one web site per installation. If you want to run statistics for more than one web site, you'll have to start over with each one—either by creating a new database from scratch or by juggling with renaming databases and the vbstat.ini file.

The setup wizard (see Figure 11-7) walks you through the process of setting up a new database for a site. It prompts you for the name of the database, the location of your server log files, the output directory, and general information about your site. When you get through the panels, just click the Finish button—the setup wizard closes, and you're done with that program.

The log file import program is called LogToDB. This program has no interface at all. When you double-click its icon to fire it off, a minimized icon shows up on your taskbar (if you're running Windows 95). The program runs, importing your log file data into the integrated Access database. When it's done, the program ends.

The next program you run is called the Restriction List Editor. Shown in Figure 11-8, the Restriction List Editor is a simple, single-panel program that allows you to apply restrictions to your report runs. Using it, you can eliminate objects (such as graphics) from your top ten lists. You can also eliminate particular Internet sites or users from your reports.

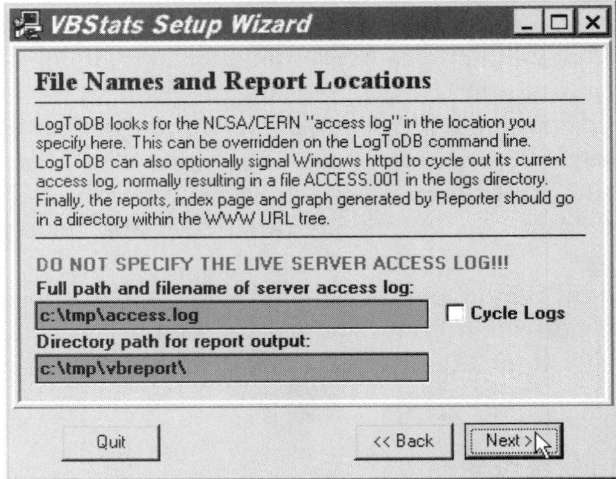

Figure 11-7. *The VBStats setup wizard gathers information from you about your site, the location of your log files, and where generated output should be placed*

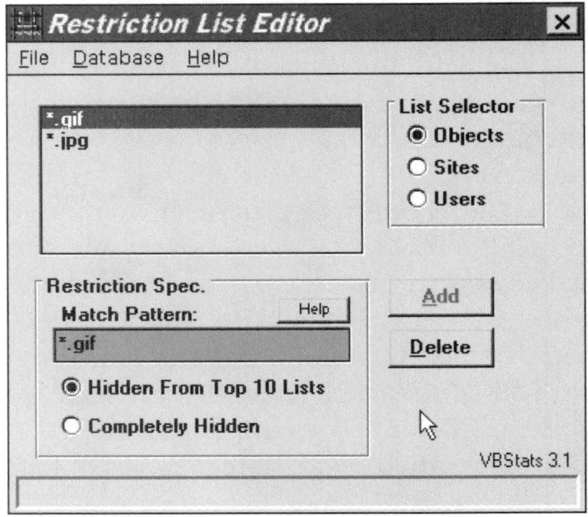

Figure 11-8. *The Restriction List Editor allows you to eliminate objects, sites, and users from your reports*

The Reporter is another interface-less module. It runs silently and generates your statistics reports in the output directory you specified.

The final program module is the Database Maintenance module, shown in Figure 11-9. This is a single-panel program you can use to clear out the contents of the VBStats database. You can choose to purge the database of records prior to a given date, clear the totals table, or compact the database (something you do after purging records).

VBStats reports are as basic as the program itself. VBStats generates an index page with a graphic depicting weekly traffic volumes and links to individual weekly reports (one for each week's worth of data in the access log you import). In Figure 11-10, you can see a bar graph showing traffic volume (in units of requests) for each week that had corresponding data in the access log. Below the graphic and under the heading "Usage Statistics Reports" are links to individual reports for each week.

The weekly reports have five parts:

- A graphic showing the daily number of requests
- Period totals for accesses (requests), data transfer volume, index queries, and the number of GET and POST methods during the period
- Top ten objects by access count
- Top ten sites by access count
- Top ten sites by byte count

Figure 11-11 shows the top of one of these weekly reports. These reports are very rudimentary. They're presented in unordered (bulleted) lists rather than in tables. The number of accesses and byte counts are not lined up because they're not in a table, and they're difficult to read because they're not presented with commas separating the numbers. Also, VBStats offers no way to configure the number of objects presented in each report. If you really need a top twenty list instead of a top ten list, you'd better look elsewhere (or be willing to play with the Visual Basic code yourself).

In summary, although it once was a very popular and highly touted statistics program, VBStats is the bottom of the barrel today. VBStats is stale, but in all fairness, the author has moved on to more interesting (and hopefully profitable) projects. However, some Visual Basic programmers are out there working on improvements, refinements, and even a 32-bit version of the program for Windows 95/NT.

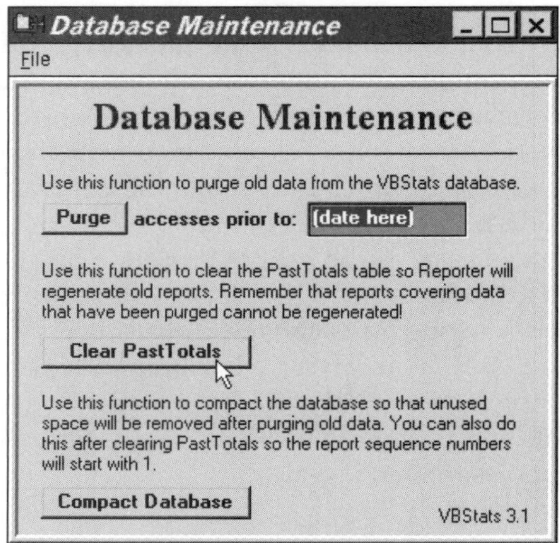

Figure 11-9. *The Database Maintenance program panel*

Other Free and Shareware Log Analysis Programs

As I mentioned in the introduction to this chapter, the field of free and shareware log analysis packages is becoming stale now that commercial developers have begun to service this market. Still, a plethora of programs that some intelligent people worked very hard on are available free—or nearly so—to you.

They're not all old and crumbling under their own weight. Some are new and even put the latest technologies to the test. One such program is 3Dstats. 3Dstats takes your server log files and generates a three-dimensional VRML model of your server load that you can view in all its 3-D glory with a VRML viewer. It's not really a serious statistics package, offering nowhere near the level of information about your site usage that current commercial packages do. It will get you to say "gee-whiz," but after you do, you will probably decide that its technology is still somewhat gratuitous at this stage.

Other gee-whiz packages use technologies like Java and JavaScript to log your site reader's activities. Of course, if you use one of these packages, either you're making the assumption that everyone visiting your site has Java capability or you're willing to throw away that portion of your visitors who don't.

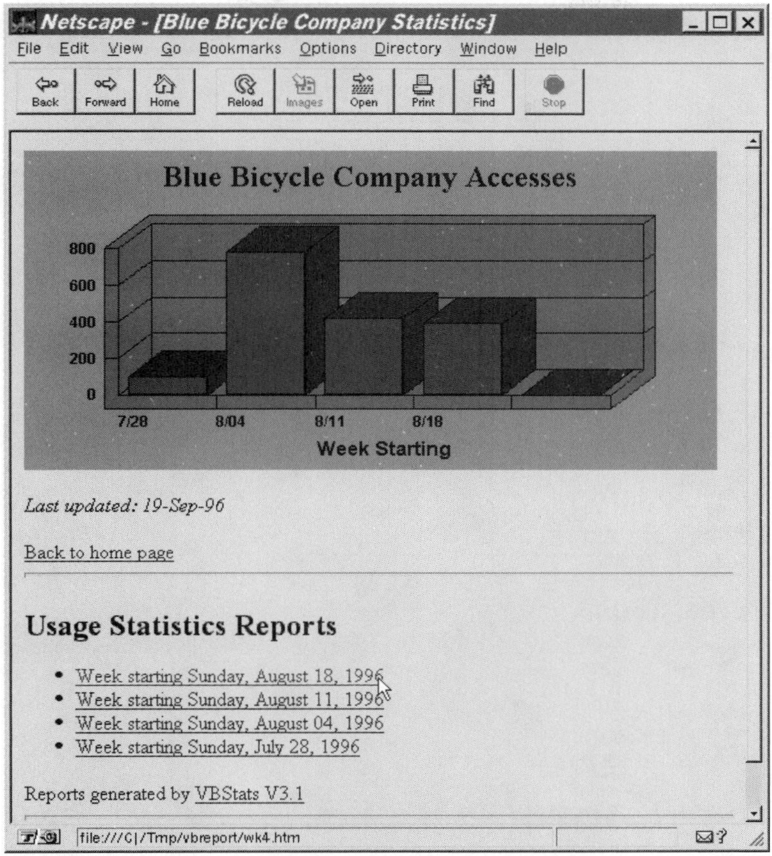

Figure 11-10. *The VBStats index page ties together the weekly reports that the program generates*

There are, however, still many more viable log analysis packages that you may indeed want to check out. To try to cover them all here would be an exercise in futility, given how quickly everything changes. If you're interested in what's out there and you've got a day to blow, check them out for yourselves. Start with the major search engines, look for "log analysis tools," and knock yourself out. Some log analysis packages are for specific servers, like Microsoft's Internet Information Server and Netscape servers. Some programs take the output from other log analysis programs to further refine their presentations and generate graphs. Some are even batch files and shell scripts to generate text-based graphs to depict your server loads.

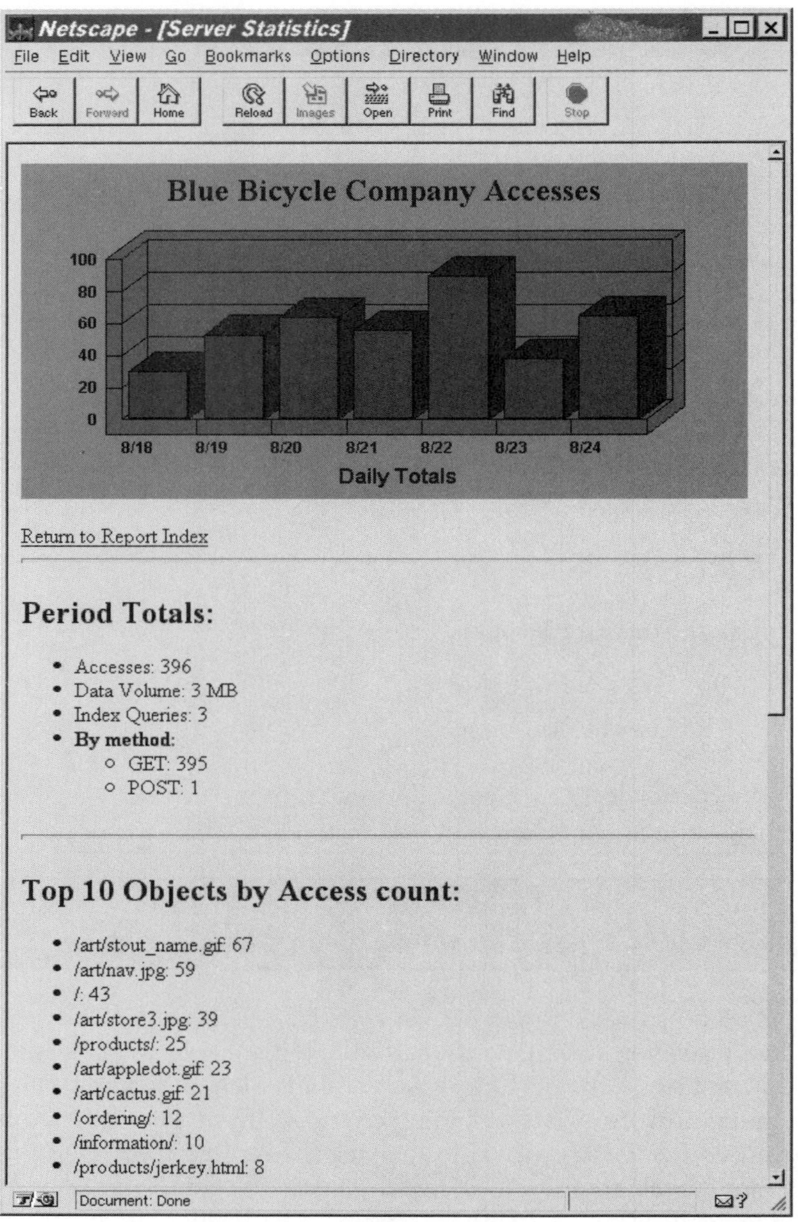

Figure 11-11. *A weekly report page*

But if you're just beginning your search for log analysis software, don't start with these. Start with the commercial developers. Nearly every one of them offers evaluation versions of their programs that you can download and try out free of charge and with no obligations. The packages that the commercial developers produce are on the cutting edge—they're aware of important features like the need to measure visits and visitors and not merely hits. They're also aware of new technologies and trends like cookies and ad servers. Once you've decided for yourself that you don't need this level of accuracy and reporting, then check out the freeware.

In the next chapter, we will take a brief look at system administration considerations that you may want to consider—things like automating log file importation and report runs, archiving, and providing reports from virtually hosted sites to customers.

COLOR
FILM

Chapter Twelve

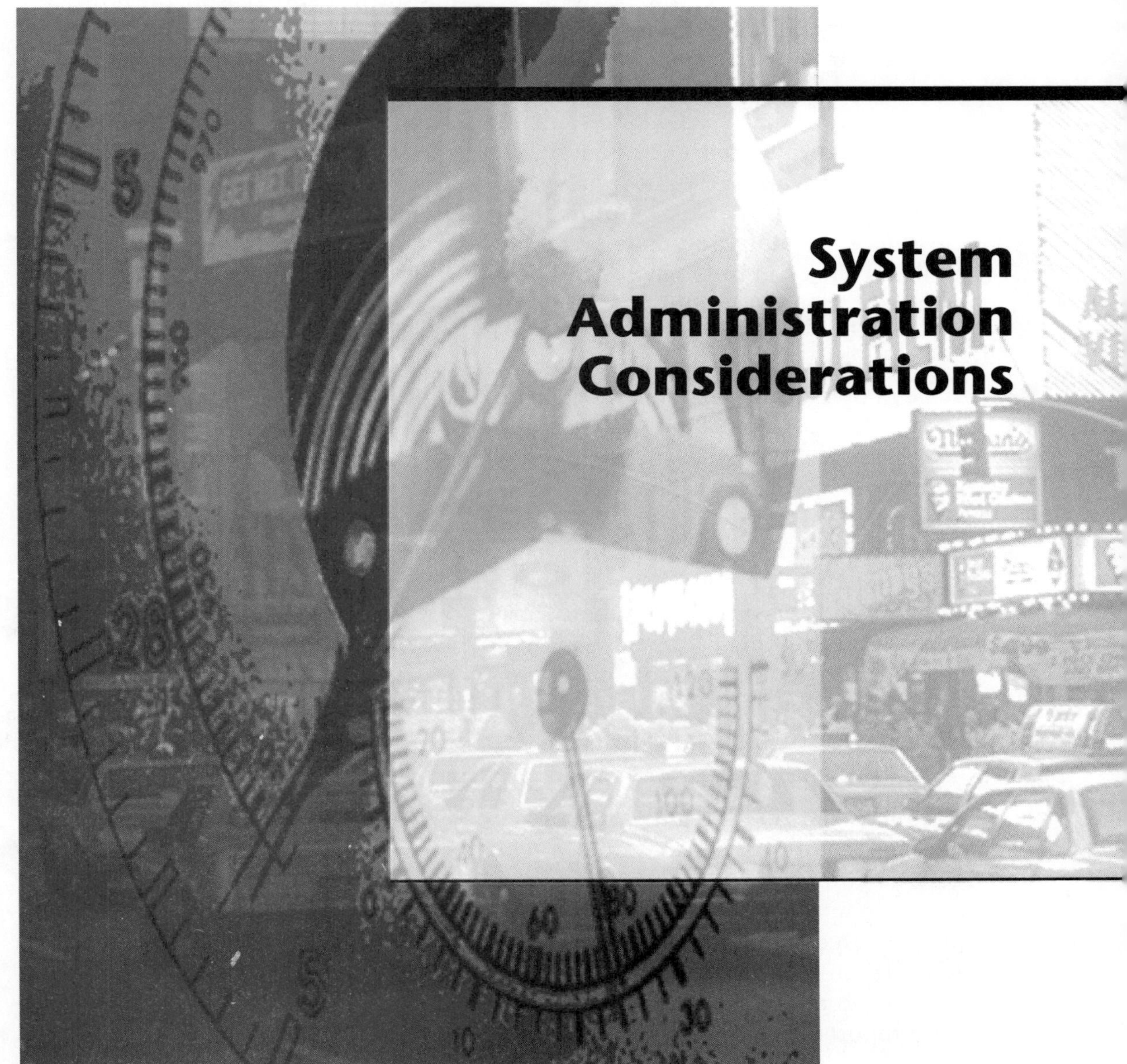

System Administration Considerations

The title of this chapter was originally going to be "Developing Your Own Statistics Package," but when it came time to write about it, the whole thing seemed downright absurd. There are just too many quality packages out there from companies with full-time programming staffs. These people work hard every day to eke out superior stats packages to edge out their competition. Moreover, the prices for statistics software are reasonable now, and they are falling all of the time as competition between these developers heats up. In short, there's just no compelling reason for system administrators or programmers to take the time to write their own statistics packages.

However, some hard-core system administrators pride themselves in doing all of their own systems administration functions—sometimes including web statistics software. For pure system administration, this is fine. Today, and probably for the foreseeable future, most of the web servers on the Internet run on Unix computers, and there's a whole bunch of system administration to be done on Unix computers without trying to reinvent the wheel by developing original log analysis software.

Even if you do choose a commercial or shareware/freeware log analysis package, you'll have plenty of system administration functions to perform. Instead of focusing on developing original log analysis tools, in this chapter we will look at some of the system administrations functions you can undertake to make your traffic measurement easier and, if appropriate, to provide your customers and clients with as much information as they want to analyze their own traffic.

In this chapter, we're not going to go into the details of scripting languages such as shell scripts and Perl or into system administration. There are too many other great books out there on these subjects. But we will take a high-level look at the kinds of things you will want to do, like automating the processes involved in accumulating and analyzing your site statistics.

Automating Site Statistic Runs

Running your server log files through a statistics package just once is an easy task to accomplish. The first time you do it, you will probably carefully take a copy of your server's transfer log and place it in a convenient location—such as in a temporary directory on your Windows computer's hard disk drive or in an FTP directory on your Unix system—to make it easy for your new software to find. You will run the software for the first time, taking note of every import option and report option, then you will begin the import process and produce your reports.

Since most of the log analysis packages create their output in HTML format, you will then have to find the resulting report files and move them to a permanent location (if you intend to keep them permanently). Or, you will rerun the entire process to change import and report parameters to suit your needs.

This is fine the first time you use your new software. It's not even that much trouble. It's as easy as the developers can make it—which is pretty easy, considering the integration of Unix web servers and (mostly) Windows analysis software. But after you go through this process more than once, you'll be looking for a way to automate the whole thing. It is more trouble than most system administrators are willing to go through, especially on a regular basis.

e.g. Software's WebTrends product has a neat feature called Schedule. Using this feature, you can schedule previously "memorized" downloads and reports to be run at the times and intervals you specify. Setting up WebTrends to run scheduled reports is easy. Figure 12-1 shows the Edit Scheduled Summary Report dialog box. You just set the parameters for the particular log file, the reports you want to run, and the date, time, and interval information and click Save. WebTrends will save your scheduled report in a list of scheduled events.

The downside to the WebTrends scheduler is that the WebTrends program has to be running and idle when the time comes for it to run the scheduled event. If the program isn't running, or if you are using it (say, generating another report), it won't catch the start time and your scheduled event won't run.

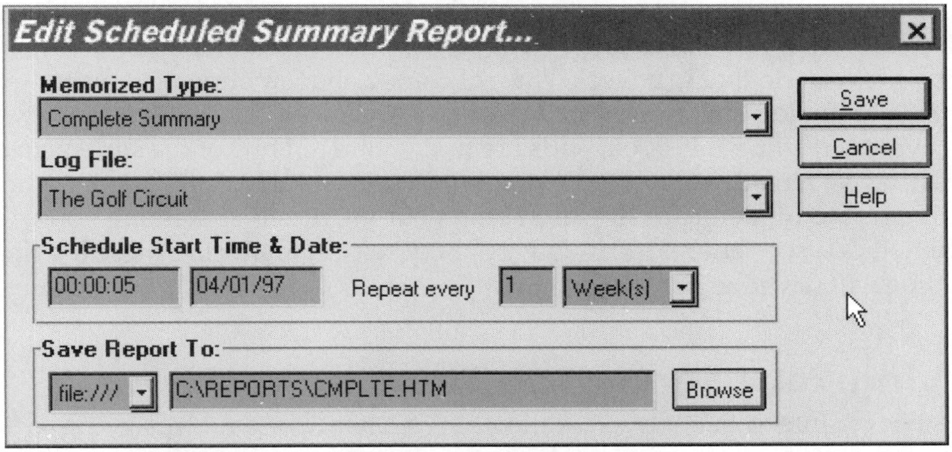

Figure 12-1. *WebTrends report scheduler*

Similarly, you can schedule reports to run with Intersé's Market Focus software. However, Intersé takes a different approach to automation—and a better one at that. Market Focus provides a command-line interface for both its import and analysis modules. Using this command-line interface and the operating system's built-in event scheduler, you can automate the importation of log file data and report running.

A system administration chore you will need to work through is coordinating your log archival process (which we will look at next) with your scheduled imports and report runs. How you do this will depend on the operating system that your web server runs under. Under Unix, you will likely use the Unix cron facility (which we will discuss in the "Implementing Automation" section). For Windows NT servers, you will likely use the NT At or WinAt program.

Automating the Archival Process

You probably realize by now that your most important log file is your transfer log. If you've really been paying attention, you know your transfer log will include your referrer and user agent data, and it will even contain your server-set cookie fields. Up to the point of the bytes transferred field, your transfer log will be in the same format as the Common Log Format, but beyond that field, it will include the additional fields of data.

Whether or not you analyze your web site statistics, you will have to set up a facility to automatically archive your log files. Especially with the addition of referrer and user agent information, your transfer log can grow big quickly. A moderately busy site can easily log 100 megabytes of data in a month. Very active sites can do that in a week. And, of course, the few superactive sites on the Net can log that much data in less than a day.

If you're reading this book, odds are that you're not one of the superactive sites (they've already solved these problems for themselves long ago). If your site is small to moderate in size, archiving your log files is something you will have to tackle sooner or later; if you are about to start analyzing those log files, the time will never be better than now.

Archiving Your Server Log Files

If you never intend to analyze your log files, archiving them is not going to be a major problem. You can just delete them once a week or once a month—whatever it takes—right from your crontab file. But then, you wouldn't be reading this if you don't intend to use those files.

All of the log analysis packages worth considering import your log data into some internal format. The Windows packages use a run-time version of a database management system like Microsoft Access or FoxPro. Others use their own proprietary formats, but they do all load up the data internally. At first blush, this may seem like all of the archiving you need. But it will become apparent fairly quickly that this won't work for long. Besides, if you need to rerun a month's statistics from scratch, you will be better off if you have the original log files to work with.

How long should you save your original log files? The answer depends on their importance. If you run a product sales-oriented site, they might not be all that important. In fact, in this case, you may choose to run your statistics reports once and just delete the source log file.

On the other hand, if you are working on building a site that will depend on ad revenue, the history of your traffic—together with the detail you used to create your reports—can be of paramount importance. In this case, why not save them all—forever? Tape cartridges and MO drive diskettes are relatively cheap. Think of the inherent value in being able to tell prospective advertisers that you have the detail log files that you used to produce every single traffic report you have ever produced.

How Often Should You Archive?

The terms that system administrators use to describe the process of archiving log files are "rolling" or "rotating" the logs. You will definitely want this to be an automated process. More than likely you will want it to happen as close to some chronological border as possible. Whether you do it daily, weekly, bimonthly, or monthly will depend on the level of maturity of your site.

When your site is new and you're monitoring all of the things that we've talked about up to this point—the error log for fixing minor problems, your cookies to make sure they're being set and logged correctly, and so on—it may be that you feel you can't go a month before seeing some traffic statistics. After your site matures just a little, you will probably want to settle into a standard of monthly or semimonthly reports.

Implementing Automation

On Unix computers, the program that allows you to make the computer do things you want it to at certain times is called cron. No doubt, the name comes from chronometer or chronology—having to do with time. Alas, the cron program is another prime example of the poor spelling skills of some grad student (this one was probably at the University of California, Berkeley, in the early seventies).

The cron program is a small and simple program that runs as a daemon. This means that it doesn't expect any input from a terminal anywhere, and it doesn't produce any output that's supposed to be viewed on a terminal screen. It just wakes up every minute and looks at a file called the crontab file to see if it is supposed to do anything besides go back to sleep.

The crontab file is a compact little text file that holds a line of data for each command that should be executed automatically by the system. Like a server log file, each line consists of a number of fields. Most of these fields have to do with telling the cron program when a particular command is supposed to execute. Of course, you should check the documentation for your particular operating system, but on most Unix systems, the fields describe the following:

- Minutes of the hour
- Hours of the day
- Days of the week
- Days of the month
- Months of the year

The final field in the crontab file is the command to execute. The entries (jobs) in a crontab file can be things as mundane as running the calendar program one minute after midnight or as exciting as daily and weekly system administration chores. In fact, your mission to rotate your log files is very similar to jobs that the system is already doing among those daily and weekly chores.

More than likely, your Unix system is already rotating some log data on your system. The number of log files that a Unix computer keeps is staggering. Log files are produced by the mail system (usually called sendmail or smail) for every FTP and telnet access and by the kernel of the operating system itself, just to name a few.

One common log file rotation scheme keeps nine backups of a log file numbered one through nine. Each time the rotation script runs, file nine is deleted, files one through eight are renamed with an incremented number, and the current log file is moved into position one. For example, with this scheme on a typical sendmail log file, a log file directory may contain the following files:

maillog
maillog.0.gz
maillog.1.gz
maillog.2.gz
maillog.3.gz
maillog.4.gz
maillog.5.gz
maillog.6.gz
maillog.7.gz
maillog.8.gz
maillog.9.gz

Assuming the mail log is set up to be rotated weekly, this scheme archives nine weeks of mail log data. If you do monthly system backups (as to tape) that you keep permanently, you will have a permanent record of your system's mail transactions.

Most Unix systems employ a log rotation scheme like this for automatically rotating their log files. This makes your job easy. All you have to do is extend it to your web server's log files or, perhaps more accurately, clone it and adapt it to your server log files, and put an entry in your crontab file for your new archiving utility.

Internet Providers and Virtually Hosted Sites

Internet Service Providers (ISPs) (or anyone hosting multiple web sites on single systems) face some challenges beyond those of the ordinary web site administrator. Many ISPs host web sites for their clients. This invariably puts them in the position of having to explain web site statistics to their customers and clients. Not only do they have to explain these statistics in great detail, they have to provide them with the tools to track the traffic on their web sites.

Most of the time, when ISPs host a web site for a client, they don't dedicate a computer to the new web site. This would be a very big waste of resources. Every web server worth using can host multiple—even dozens, if not hundreds—of web sites on a single computer. The limitations are the speed and the processing power of the computer and the network throughput. In the industry, this is called *virtual hosting,* or virtual site hosting.

We won't go into the details of how to set up a system to host multiple sites here; most server documentation covers that adequately. But we will discuss the impact of maintaining server log files for multiple hosts.

Whether a site administrator or webmaster works on his or her own computer, the issues surrounding traffic statistics are the same. Site owners want to know who and how people are using their site. Most often, the hosting services that ISPs offer don't include any comprehensive analysis of site statistics. This would be a tough nut for ISPs to crack.

Regardless of what log analysis software an ISP chose, it wouldn't be right for all of its customers. Inevitably, some clients would have needs and preferences that differed from those of others. As a result, most ISPs either provide their clients access to the raw server log files for their sites or provide them with automatically generated summary statistics of the traffic at their site.

For many small web site owners, summary statistics (like those shown in Figure 12-2) are adequate. They simply want to know if people are accessing their pages and, if so, which ones and how many.

Other web site owners will want more detail. ISPs can satisfy these customers by configuring their web servers to keep separate server log files for each one of the sites that they virtually host. The ISP can still do any processing that it has to do on the client's log files, like calculating bandwidth usage. Then, with a daily, weekly, or monthly script, the server can automatically rotate the logs into a directory on the system that the client can access. This way, the client can have its own log files and use whatever log analysis software or service it likes.

When you set up a web server to serve multiple hosts, one of the things you have to configure is the server's log files. Of course, you should check your server documentation for the specifics for your server, but in general, you just give the server a path and filename for the particular virtual host. On the free Unix servers, this is in the server configuration file—normally called httpd.conf. In the httpd.conf file on these servers, you can specify a different transfer log for each virtual host with the TransferLog directive.

There wasn't a great level of detail in this chapter about these system administration issues, but that wasn't my goal. Instead, I wanted to give you

```
Daily Access Report
Accesses    Size     Document
============================================================
       1     10k /art/acknowledge.jpg
       1      6k /art/authors.jpg
       1      7k /art/contents.jpg
       2      1k /art/flask.gif
       1      9k /art/howtouse.jpg
       2     14k /art/logo.jpg
       2      3k /art/million.gif
       2     23k /art/nav.jpg
       2      4k /art/observatory.gif
       1      6k /art/rick.jpg
       1      1k /art/rocket.gif
       2     13k /art/welcome.jpg
       1      0k /authors
       1      2k /authors/index.html
       1      0k /authors/rick
       1      2k /authors/rick/index.html
       1     20k /authors/rick/rick.jpg
       1      0k /contents
       1      5k /contents/acknowl.html
       1      6k /contents/howToUse.html
       1      1k /contents/index.html
       1      0k /welcome
       1      5k /welcome/index.html
=============================================================
      29    138k Date: 08/25/96
```

Figure 12-2. *A simple daily access report supplied by an ISP*

a high-level overview of the kinds of system administration chores you will want to think about as you implement and automate the generation of your traffic statistics. I hope I've accomplished that. In the next chapter (and the next section) we will jump into advertising and marketing issues, and how these impact, and are impacted by, your traffic analysis.

COLOR

PART FOUR

Marketing and Advertising

Chapter Thirteen

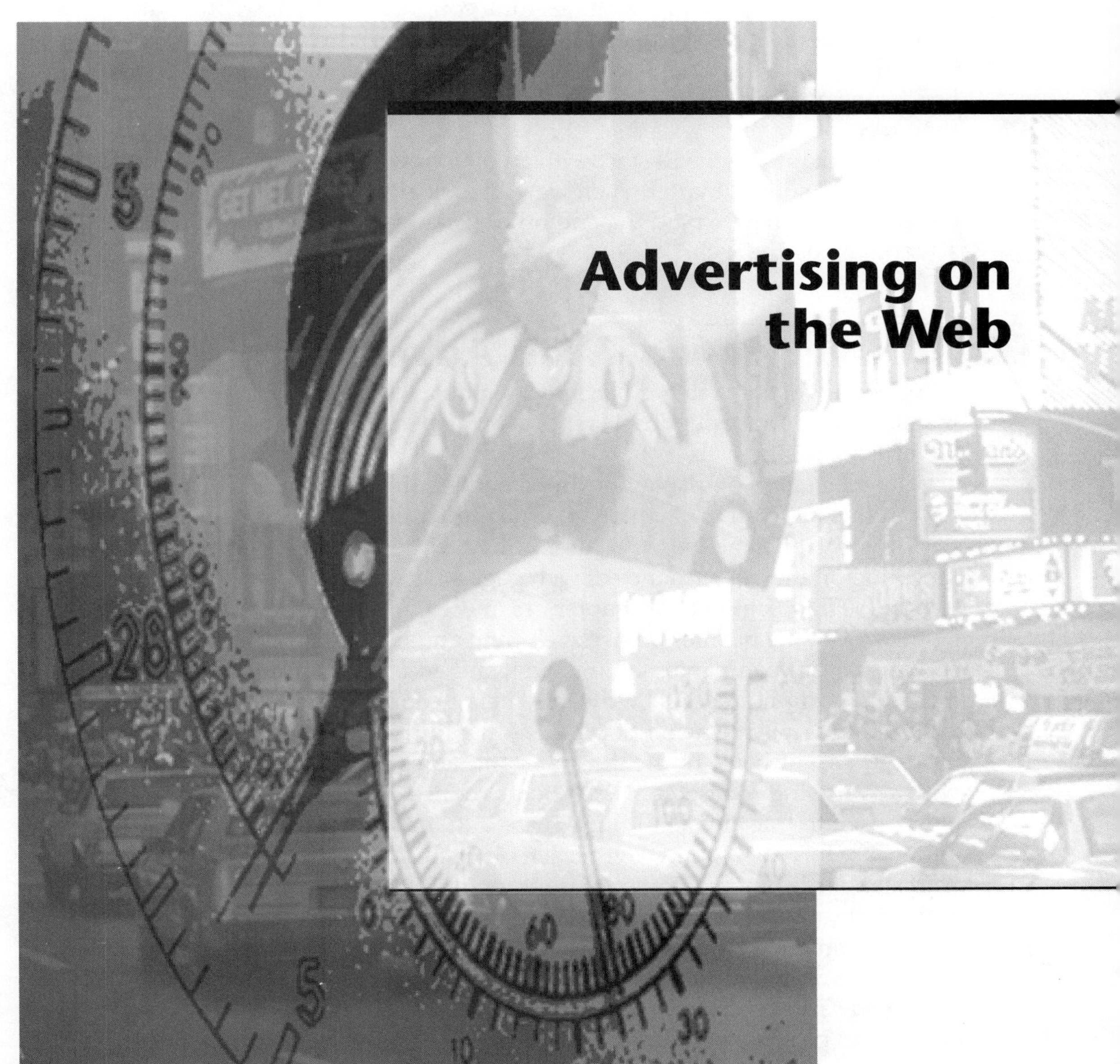

Advertising on the Web

In this part, we come to what will be the meat of the book for many: advertising and tracking the effectiveness of advertising on a web site. We will look at how to run advertisements on your site and we'll look at software and services to track the effectiveness of ads and ad campaigns.

Web-based advertising isn't really new. We have all been seeing banner ads on web sites and search engines for a couple of years now, but that view of advertising is completely superficial and isn't really the aspect of advertising that is germane to this book. In the next four chapters, we will be looking at what goes on behind the scenes—and there's a whole lot more than meets the eye.

In this chapter, we will start with an overview of this very new and very complicated subject matter. Then we will look at what makes banner ads effective, how they're paid for, and other general topics of interest relating to advertising on the web.

An Overview of Web Advertising

Advertising on the Web has become a big business. It seems that every week a magazine or newspaper article reports on a new study or finding about advertising on the web.

The mechanics of displaying an ad banner are really quite simple (we will look at this in detail in Chapter 14). The mechanics of reporting on the performance of an ad isn't too complicated, either—in fact, it's very similar to reporting on the performance of web pages in general, with a couple of minor twists. We'll look at this in Chapters 15 and 16.

However, the businesses of developing and marketing ad servers, ad serving service companies, ad tracking, and ad performance auditing are incredibly complicated. As I write this, literally dozens of new, small companies each thinks it has the next hot product. In addition to these, most of the larger, more mature web site traffic tracking companies are getting into this niche with advertising-focused services. Some of these companies are forging alliances with traditional auditing firms and media measurement companies. Others are forging alliances with each other through mergers and buyouts. And standards organizations are popping up like weeds, all vying to define and set the standards. In short, this industry is currently a huge, tangled web of confusion and complexity.

Right now, nobody knows how it will all settle out. There are companies selling ad server software (NetGravity and ClickOver, for example), companies promoting their ad networks and giving away their software

(Real Media), and companies tracking ad activity from their own sites (Focalink, NetCount, and I/PRO). We don't even have any common standards in pricing yet: some of these companies are charging flat monthly fees; others have a tiered or structured rate schedule; still others take a cut of the ad revenue paid by advertisers.

We'll see how this will all play out over the next year or so. In all likelihood, we will end up with a handful of companies that offer similar but slightly different services and that charge for their services in a variety of ways.

Ad Issues

Let's look at some of the issues facing ad publishers and advertisers, including the effectiveness of ads, the rates advertisers pay for web ad space, and how they like to pay for it.

Ad Effectiveness

Most advertisers, site owners, and related companies offering services and selling software generally agree that the effectiveness of an ad banner depends on four primary criteria:

- The relevance to the reader's interests
- The quality of the banner graphic itself
- The number of times readers have seen the banner
- The placement of ad banners on a page

The relevance of the ad to the reader's interests is arguably the most important factor. This is called *targeting.* For example, advertising sports equipment on a sports-related web site will generally be a pretty good fit; similarly, pharmaceutical supplies would be well targeted on a page geared to physicians and others in the health care industry. On the other hand, an ad for carpet cleaning on a page devoted to motorcycle enthusiasts wouldn't be a particularly good fit (well, maybe if they work on their motorcycles in the living room). Some products—pillows, for example—are used by nearly everybody, but would an ad for pillows and comforters be particularly good on any web site? Probably not.

The quality of banner ads is crucial. Even beyond the technical quality is the *magic* of the ad, which no one can predict. Two quality ads for the same product or company served in rotation on the same page can have dramatically different click-through rates. Sometimes it's just impossible to predict which ones will work and which ones won't. This is why it's always a good idea to create a collection of ads (which are often referred to in the industry as *creatives*) and test them to see which are the most effective at generating click-throughs.

Numerous studies have shown that the effectiveness of ads degenerates quickly in relationship to the number of times a reader has seen an ad. If they haven't clicked on the ad after the third time they've seen it, chances are less likely that they ever will. In late 1996, I/PRO and DoubleClick (the largest ad network) teamed up to do a comprehensive study of ad banner impressions and click-throughs. In their study (titled "The Web in Perspective: A Comprehensive Analysis of Ad Responses"), I/PRO and DoubleClick report the details of their findings regarding the relationship between precise reach/frequency and user response to ad banners.

Regarding ad placement, ad banners are usually rectangular in shape and are placed at the top of a page. Studies have shown, however, that varying the size, shape, and location can also affect the performance of an ad. An example of one variation is an ad shaped like a medallion and placed in the middle of a page, perhaps with the text of a page wrapped around it. This kind of variation can dramatically increase the performance of an ad.

Ad Rates

Traditionally (that is, in traditional media), advertising rates are based on the number of *impressions* an ad can make—in other words, the estimated number of people who will see an advertisement. In some media, like magazines, the number of impressions may actually be more than the number of copies of the magazine that are mailed out and sold on newsstands. This is due to an imaginary multiplier called a *pass-along* rate, which refers to an assumption publishers like to make that some publications will be passed along to additional people, who will also have the opportunity to see an ad.

Web advertising has no pass-along ratio (it is a rare case indeed in which a second reader is looking over the shoulder of someone browsing the web). However, the term "impression" has been adapted to web-based advertising, where its meaning is taken as a reader seeing an ad banner—or at least seeing the page on which an ad banner is displayed.

Impressions vs. Click-Throughs

The biggest issue today facing web advertisers, site owners, and ad companies are how ads should be paid for. Site owners (the publishers) want to be paid for ad space by the impression, which is the way other media charge for ad space. Site owners love to point out that they aren't responsible for the effectiveness of an ad—they're only supposed to supply the readers, and that's how they want to be paid.

Advertisers, on the other hand, realize that in the web they have a new medium with which they can do something they've never been able to do before—directly measure the effectiveness of an ad (well, maybe not quite yet, but we're getting there). Right now, we can measure fairly accurately the number of impressions an ad makes, and that's much more information than any other medium can provide.

Besides measuring impressions, we can also measure the number of people who click through an ad banner to go to the advertiser's site. This is like having a surefire way to tell if people bought a particular brand of soap because they liked the ad they saw for it on TV. (Marketing types get very excited about the prospects of this information.)

It's true that clicking through an ad banner and going to an advertiser's site is still a long way from buying a product on that advertiser's site. But getting readers there is the first big challenge that advertisers have to tackle, and being able to test the actual effectiveness of the ads that draw readers is an alluring notion.

Because it's technologically feasible (downright easy, in fact) to measure ad banner click-throughs, advertisers would like to shift the industry norm payment structure from impressions to click-throughs. For example, instead of paying less than a penny for an impression, some advertisers would like to pay a higher rate—say, a nickel—for each click-through.

Here's where the publishers (the site owners selling the ad space) throw out the penalty flag, asking how they can possibly be held responsible for the quality of the ad banner and the advertiser's campaign. For example, two similar ads for the same product or service displayed on the same page can have significantly different click-through rates; in such a case, it's clear that the difference in performance can only be attributable to the difference in the ads.

Many industry insiders speculate that in the end we will find a combined rate structure that takes into consideration both impressions and click-throughs. For example, we might see impression rates (say, a penny per impression) plus compensation for each click-through at a higher rate (say, a nickel). This still places some of the burden of an ad's success or failure on the web site owner—for which they didn't have any responsibility before—but the payment for click-throughs is payment above and beyond what they get now. Under a fee structure like this, the web site owner has an incentive to work closely with advertisers to help them determine which ads are the most effective and to give them the placement on pages they need to maximize click-throughs.

Impression Rates and Targeting

Most sites and advertisers still agree that compensation should be based on the impression. Actually, the common way to express an ad rate is by the cost (in dollars) per thousand impressions. This is usually abbreviated as CPM (cost per M—with M being the Roman numeral for thousand).

Typical impression rates as I write this can range anywhere from $20 to $100 CPM. That's two cents to a dime per impression. The wide fluctuation in price range depends on how narrowly focused the targeting is. Let's look for a minute at targeting.

Ad servers and ad server companies (and ad server networks) can use the incoming HTTP header information from readers to determine all of the usual statistics about a reader. This information includes the following:

- The domain name of the reader's computer
- The type of web browser the reader is using
- The operating system run by the reader's computer

Some ad servers and ad networks (DoubleClick, for example) have extensive databases from which they can derive a tremendous amount of information from the domain name alone. For example, from any domain name, DoubleClick can determine the following:

- The name of the company or organization
- The geographic location of the organization
- The SIC code (i.e., the type of business), if applicable
- The size of the company

This results in a typical targeting suite of the following items:

- The site (or computer) a reader is coming from
- The frequency of visits
- Geography

- Domain name
- SIC codes
- Company size
- Browser type
- Operating system
- The Internet Service Provider (ISP)

One targeting criterion, however, isn't included in the above list: the subject or the content of the site or sites running an ad and the relevance of the ad to that content. Remember that this remains the most important criterion. Remember also that individual site owners can always contract directly with an advertiser. Site owners and advertisers can work together to make sure that the audience of a site is the appropriate target for the advertiser's products or services.

A completely untargeted campaign is one where none of the above criteria is used to narrow the focus of ad views. For example, if Coca-Cola advertised on the web, it probably wouldn't be too interested in restricting its ads to certain companies, company sizes, geographic locations, or specific browser users. Any person browsing through web sites is a potential Coke drinker, so there's no inherent advantage to restricting ad views on any criteria.

Completely untargeted campaigns typically draw the lowest rates; $20 to $25 CPM (or 2 to 2.5 cents per impression) are the norm. From there, using targeting criteria usually elevates rates dramatically. For example, Figure 13-1 shows the DoubleClick network's rate card (as of 1996). As you can see, the lowest rate visible on the card is $25 CPM—the "base" rate for the DoubleClick network (the top row of the table). This is the rate for a completely untargeted campaign placing ads into all of the member sites on the network.

Targeting an ad campaign to specific sites is going to cost more. The next six rows of the table in Figure 13-1 represent the CPM for some specific sites that belong to the DoubleClick network. These are premium sites with very high traffic offering a range of targeting focuses.

The last five rows of the table show the categories of affinity sites in the DoubleClick network. An *affinity site* is a special-interest site (like our golf site). By choosing to target to these sites, the advertiser can narrow the focus of an ad campaign to people interested in specific topics. On average, affinity sites garner a higher rate than an untargeted shotgun approach, because there are always product manufacturers, service providers, or software developers of some sort that make products for people interested in nearly every hobby, interest, or pursuit. If these companies are at all interested in advertising on

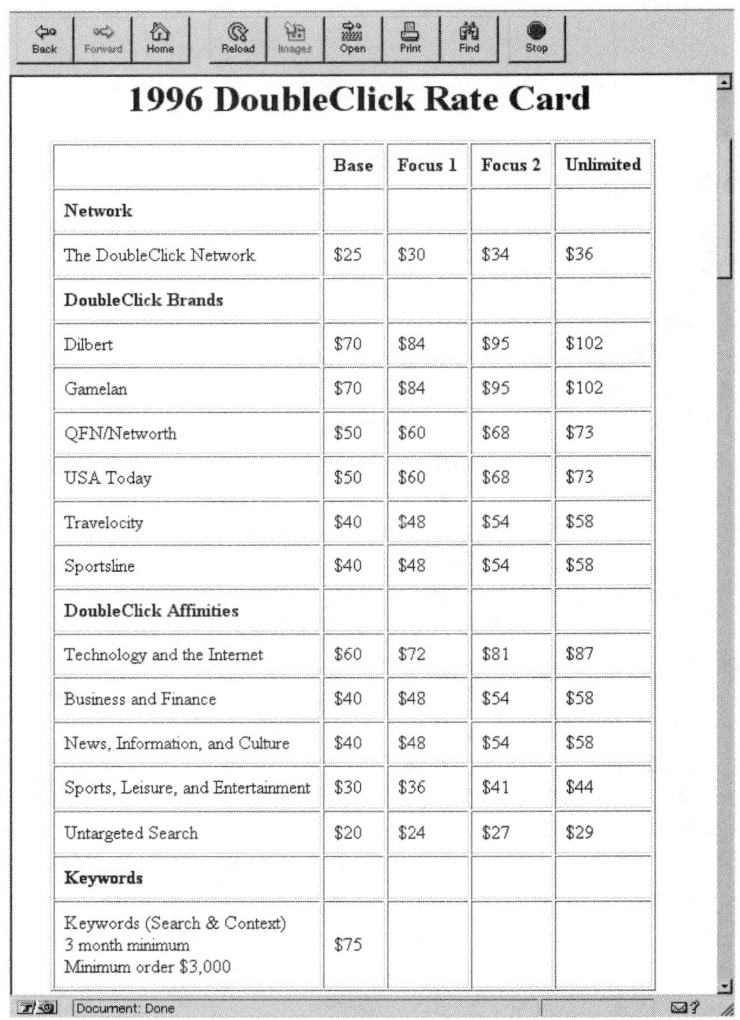

1996 DoubleClick Rate Card

	Base	Focus 1	Focus 2	Unlimited
Network				
The DoubleClick Network	$25	$30	$34	$36
DoubleClick Brands				
Dilbert	$70	$84	$95	$102
Gamelan	$70	$84	$95	$102
QFN/Networth	$50	$60	$68	$73
USA Today	$50	$60	$68	$73
Travelocity	$40	$48	$54	$58
Sportsline	$40	$48	$54	$58
DoubleClick Affinities				
Technology and the Internet	$60	$72	$81	$87
Business and Finance	$40	$48	$54	$58
News, Information, and Culture	$40	$48	$54	$58
Sports, Leisure, and Entertainment	$30	$36	$41	$44
Untargeted Search	$20	$24	$27	$29
Keywords				
Keywords (Search & Context) 3 month minimum Minimum order $3,000	$75			

Figure 13-1. *The DoubleClick Rate Card for 1996*

the web, they will surely prefer to do it on a site devoted to the subject matter of their products or services.

So the first important criterion I mentioned in the "Ad Effectiveness" section—choosing sites with relevant content—is addressed with DoubleClick by choosing from the rows of their rate table. Of course, this is from an advertiser's perspective. If you're a publisher (a site owner), your first goal should be to have a premium-quality, highly trafficked site (but of course, you already know that).

Returning to Figure 13-1, the columns of the rate card table represent the number of additional targeting criteria you employ. In DoubleClick's nomenclature, "Focus 1" means that you employ one of the targeting criteria that we listed earlier in the chapter (company size, geographic location, etc.). The "Focus 2" rate applies if you employ two targeting criteria; "Unlimited" applies if you employ three or more targeting criteria.

For example, to run an ad campaign that inserts ads into the Gamelan site (a very popular site featuring the latest and hottest Java applets) with three or more targeting criteria will cost you $102 CPM, or 10.2 cents per impression. To insert an ad into one of the business and finance-related sites with one targeting criterion costs $48 CPM, or just under a nickel per impression.

Let's turn it around now and consider how much money you can make if you're the publisher, or site owner. Until you make it really big and get onto the premium lists of ad networks like DoubleClick, you're going to have to live with lower rates—around $20 or $30 CPM.

However, if you can develop a popular affinity site, your chances are good for a slightly higher rate because of the focus of your site. As I mentioned earlier in this section, there are manufacturers and service companies that sell products into nearly every market; for them, choosing a site with a focus that parallels their products is worth the extra money.

Let's look at a hypothetical case of a special-interest site—say we develop a site devoted to women's basketball. Such a site could draw interest for advertisements by sports clothing and equipment managers, among others. Let's also say that we work hard on the site to draw the traffic. The minimum amount of traffic to be considered by most advertisers and ad networks is around 100,000 impressions per month. If we can sell all 100,000 impressions and get an average rate of, say, $30 CPM, we will make the following:

100,000 impressions × $.030 = $3,000 per month

This amount is for the bare minimum number of impressions that we can offer just to draw any interest from advertisers. What if we can serve 10,000 impressions per day? That's $300 per day, or around $9,000 per month. Of course, the really big sites serve many more impressions than this, and their abilities to offer precise targeting ensure that they will get better rates as well (but they're probably not reading this book anyway).

Can You Make Any Money Selling Advertising on the Web?

This is the big question on everybody's mind. Yes, you can make money, but you've got to overcome two big obstacles. First, you have to draw enough traffic to interest an advertiser. Second, you've got to find the advertisers and sell them the ad space. The keys are the focus of your site and the amount of traffic you can draw. The focus of your site is probably what it is already (unless you're only now beginning to contemplate it). Drawing traffic is the hard part. It's easy to talk about but not so easy to do.

Chapter 14 will deal with the mechanics of serving ad banners yourself for your own site or for a small ad network. In Chapters 15 and 16, we will look at ad networks, ad server software, and ad tracking and auditing services.

Chapter Fourteen

Web Ad Mechanics

At the very core of web-based advertising is the banner ad. At its simplest, a banner ad is just another inline image embedded in a page as a hypertext link. But there's much more that you can do to make your banner ads smart, even providing you with up-to-the-minute information about how many people are seeing your ads and how effective the ads are.

In this chapter, you will learn about the mechanics of simple banner ads and about ways to log and measure their effectiveness. You will also see some technology and techniques that you can use to enhance both the effectiveness and numbers of people seeing your ads.

The Simple Banner Ad

We've all seen banner ads like the one shown in Figure 14-1. You can't visit Yahoo or any other popular site on the web without encountering them.

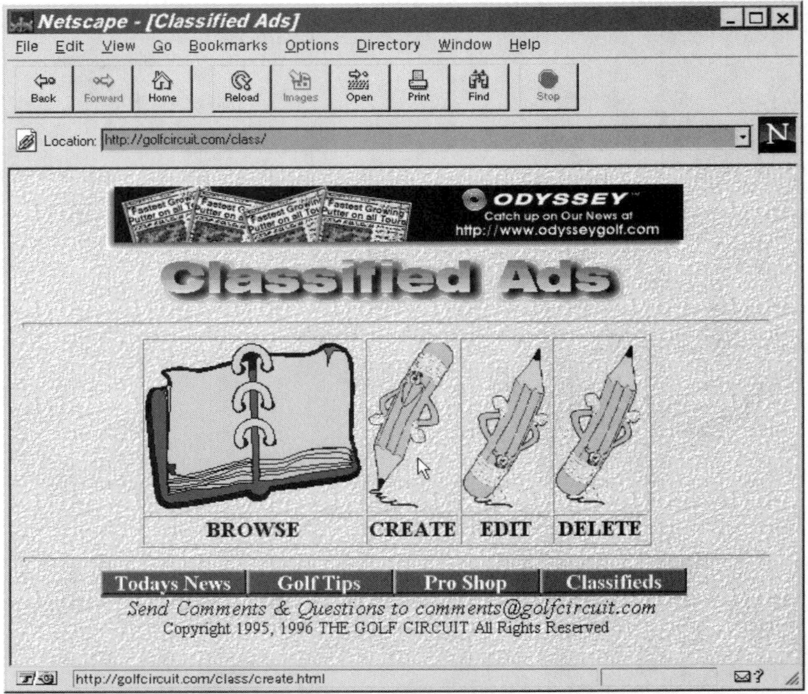

Figure 14-1. *A simple banner ad*

Whether or not you've ever been motivated to click on one, you've been affected by them.

In its simplest form, a banner ad is just a graphic image displayed on a web page. Most web advertisers have their own sites, so most often their banner ads are hyperlinked to their own site, but this isn't necessarily a requirement. Some advertising, especially a type known as institutional advertising, isn't really designed to create any immediate response. In *institutional advertising,* an ad is intended more to deliver a message than merely promote a product—or, at least, any particular company's product.

For example, we all probably remember the "Everybody (or every body) needs milk" campaign from television. This ad campaign was highly effective in delivering the message that milk is good for you (whether or not it's true). This campaign was sponsored not by a single dairy or food-processing company but by the dairy industry as a whole. Sure, the organization that produced and paid for the ad was composed of member dairies, but the ad campaign promoted the major product of the entire industry—whether everyone in the industry contributed money or not.

For the most part, the reality of advertising on the web today is that banners do lead to their sponsor's pages and they do promote products or services. Still, not every ad is necessarily intended to produce an immediate response. With some advertisers, it's the impression the ad leaves with readers that is the important aspect. But enough of philosophy. Let's get into the mechanics of ad banners.

You've probably embedded graphics in a web page. You've probably also made graphic images link to other graphics or other pages. Then you surely know that you display a graphic with the <IMG> tag and you make an image a link to another with a regular anchor tag. Here's the tag to embed a graphic:

```
<img src="banner.jpg">
```

To make this entire image a hypertext link, you just enclose the whole thing in an anchor tag:

```
<a href="http://mysite.com">
<img src="banner.jpg">
</a>
```

Or, of course, you do the following:

```
<a href="http://mysite.com"><img src="banner.jpg"></a>
```

Elementary HTML, right? Yes, but it's about to get interesting.

You can display an ad banner on your web site this way, but there's a problem: the only way you can know if the ad is being seen is by running your web site traffic statistics on your server transfer log and looking at the number of page views for the page with the ad on it. (Or, you could do a "tail -f" on the actual transfer log and visually watch the hits and read the request lines as you look for views of this particular page.)

The Poor Webmaster's Ad Tracking System

Without any expense and without hurting your brain too much, you can implement a much smarter banner ad. Using the tricks and techniques I will show you in this section, you can put banner ads on your pages that record an event every time they're viewed and clicked through.

To really seriously track ad exposures and click-throughs, you should check into ad server service providers and ad networks; we'll see a lot more about these in Chapters 15 and 16. Using the Perl scripts and methods in this section, however, you can create your own "poor webmaster's" ad tracking system.

This isn't much more complicated than a simple banner graphic displayed inline as an HTML hyperlink. Recall that a simple banner ad is composed of two parts in HTML: the tag for the image itself (the <IMG> tag) and the tag that sandwiches the image and makes the image a hyperlink (the <A> tag).

The goal of our tricks here is twofold:

- To record views of the image itself
- To record click-throughs on the image

Let's look first at how we would record a view on the image itself. Let's start with the HTML for the simple banner ad:

```
<a href="http://mysite.com">
<img src="banner.gif">
</a>
```

All three lines of code here are part of an anchor (<A>) tag. The top line defines the href for the anchor, and the bottom line terminates the anchor tag. The line in between, of course, is the tag for the image, which will become a hypertext hot spot because it's sandwiched in the anchor tag.

Tracking Impressions with Perl

Since we're going to look first at how to record the view on the image itself, we'll be working first with the middle part of the anchor—the tag for the image.

The first thing to realize is that we're going to need an executing program to open a log file and write to it. The tool we have available to do this with web servers is CGI—the common gateway interface. So, rather than referencing the graphic file directly, we're going to reference a CGI script that will do the logging we want to do and then display the graphic image file.

The tag sandwiched inside of the anchor tag needs to remain an <IMG> tag so that the browser rendering the page will know to expect an image when it is requested. The source for the image is going to be our new executable CGI script rather than the filename of the graphic file. However, to make as general a script as possible (one that can be used with many different banner ads and sites), we should pass some information into the program when we call it from an HTML file. What the script needs most is a name from which it can construct a log file name and the path and name of the graphic file we want it to display.

Ideally, we'd like to reference the graphic in the HTML like this:

```
<a href="http://mysite.com">
<img src="/cgi-bin/logview.pl?cidermill+/art/banner3.gif">
</a>
```

Remember, the anchor doesn't get invoked until a reader clicks on the image, but the image source is invoked immediately when a reader's browser displays the page. In this example, the browser makes a request for the CGI program /cgi-bin/logview.pl. The server recognizes the request for the executable program, fires it off, and passes the program its two parameters (cidermill, and /art/banner3.gif).

Figure 14-2 shows the source code for the Perl program called logview.pl. This program accepts the two arguments from the web page (a name with which to create a log file and the filename of the graphic image file to display). The logging takes place in the middle of the script. First, it checks to see if the file already exists; if not, it creates the file. Then the script opens the file in append mode and writes a time stamp and the IP address of the reader to the file.

The last block of code in the file sends a response header (the content-type header) to the browser so that it knows the following stream of bytes will be an image. Then, it sends the image. For simplicity, we've made this script assume that the graphic file it will be displaying is a GIF image. In practice,

```
#! /usr/local/bin/perl
# LOGVIEW - Logs banner views
#
# Called from an HTML link of the form:
# <a href=/cgi-bin/logview.pl?LOGNAME+http://url.com><img bannerimage.gif></a>

$docroot = '/var/www/docs/rlsnet';
$name    = $ARGV[0]; # name of the link (for the log file)
$gfile   = $ARGV[1]; # filename for the graphic to display
$logfile = "$docroot/adlogs/$name"."_views.log";
$picfile = "$docroot/$gfile";

($sec,$min,$hour,$mday,$mon,$year,$wday,$yday,$isdst) =
    localtime(time);
$tstmp = sprintf("%02d/%02d/%02d:%02d:%02d:%02d",
    $year,$mon,$mday,$hour,$min,$sec);
if (open (fH, ">>$logfile")) {   # log the view to the log file
  print fH $tstmp, ," ", $ENV{'REMOTE_ADDR'}, "\n";
  close fH;
}

if (open(fH, "$picfile")) {      # output the graphic image
  print "Content-type: image/gif\n\n";
  while (read(fH, $foo, 1024)) {
    print $foo;
  }
  close fH;
}
```

Figure 14-2. *The source code for logview.pl; this script records a view (or impression) on a page with an ad and logs the event in a log file*

you will probably want to test the filename extension of the graphic file and send the appropriate content-type header (e.g., image/gif, image/jpeg) based on the filename.

With this code in place, you've got all you need for quick-and-dirty ad impression tracking. This Perl script will record the date, time, and IP address of remote readers for every view on this page. Bear in mind, though, that while this script will give you a rough idea of how many people are seeing your ad banners, it doesn't take into account things like canceled transfers or cached pages. But it will give you a good estimate of new impressions on each ad without the need to view your server's log files or to run your log analysis software every day to see what's happening.

Tracking Click-Throughs with Perl

Just as you can log ad impressions with a Perl CGI script, you can log click-throughs. In concept, the script to log click-throughs is even simpler than the one to log impressions. Recall that after we implemented the script in the previous section to record impressions, the HTML tags to implement it were the following:

```
<a href="http://mysite.com">
<img src="/cgi-bin/logview.pl?cidermill+/art/banner3.gif">
</a>
```

Our new script to log click-throughs won't affect the image tag in this HTML (we've already taken care of that), but it will change the surrounding anchor tag. Instead of linking directly to the ad's target site, link it to another CGI script and pass it the URL of the target site (along with a name from which it should generate a log file for the click-throughs).

To remain consistent with the order of parameters from the prior script, let's say that a call to this script should look like this:

```
<a href="/cgi-bin/logclick.pl?cidermill+http://cidermill.com">
<img src="/cgi-bin/logview.pl?cidermill+/art/banner3.gif">
</a>
```

Like the first script, this one takes two arguments: a name (as for a site or ad campaign) that the script can use to construct a filename for the log file it will be writing to and the URL, which is the ultimate destination at which the reader will arrive after clicking on the graphic. Figure 14-3 lists the source code for this script, which is called logclick.pl.

As you can see, logclick.pl is very similar to logview.pl. Both take two arguments, derive a filename for the log file, generate a time stamp, create the log file if necessary, and log the event in the log file. The major difference is in the last block of code. Where the logview.pl script outputs the contents of a graphic file to the reader's browser, logclick.pl simply redirects the reader's browser to the destination URL.

Summarizing Ad Banner Log Data

Both logview.pl and logclick.pl log their respective events with a time stamp, a space character, and the IP address of the reader's computer viewing or clicking through an ad banner. Here's what the log file looks like:

```
96/08/24:15:21:46 198.68.174.83
96/08/24:15:22:29 204.216.152.7
96/08/24:15:25:03 212.186.64.61
96/08/24:16:01:42 192.188.121.87
96/08/25:05:36:51 191.82.124.5
```

```
#! /usr/local/bin/perl
# LOGCLICK - Logs banner click-throughs
#
# Called from an HTML link of the form:
# <a href=/cgi-bin/logclick.pl?LOGNAME+http://url.com><img bannerimage.gif></a>

$docroot = '/var/www/docs/rlsnet';
$name    = $ARGV[0]; # the name for the log file
$url     = $ARGV[1]; # the target destination
$file    = "$docroot/adlogs/$name"."_clicks.log";

($sec,$min,$hour,$mday,$mon,$year,$wday,$yday,$isdst) = localtime(time);
$tstmp = sprintf("%02d/%02d/%02d:%02d:%02d:%02d",$year,$mon,$mday,$hour,$min,$sec);

if (open (fH, ">>$file")) {           # log the click to the log file
  print fH $tstmp, ," ", $ENV{'REMOTE_ADDR'}, "\n";
  close fH;
}

print "Location: $url\n\n";           # redirect to the destination
```

Figure 14-3. *Logclick.pl records clicks on a banner ad in a log file, then directs the reader's browser to the banner's target page*

Using the combination of an administrative web page (one that only you as administrator or webmaster can get to) and another CGI script, you can make it easy to see up-to-the-minute statistics on your ad campaign anytime you want.

The first step is to create this administrative web page. It's probably a good idea to put such a page in a secure area of your web site. If you don't already have one, you might consider creating a /admin directory on your site and password-protecting it with basic user authentication. Virtually every server you would want to use allows you to do this. Check the documentation for your particular server to find out how.

Figure 14-4 shows the Ad Campaigns page I created to monitor ad banner activity. Actually, this is an intro page, where you pick the ad campaign that you want to check up on or monitor. There's no magic here—this is just a simple form with radio buttons of the same name and "values" that represent the name of each campaign from which the script that processes the form will construct filenames for each choice. The field is named campaign, and here the "value" of the field for the selected option is cidermill. When I click on the Monitor button, the name/value pair campaign&cidermill is passed to the CGI script.

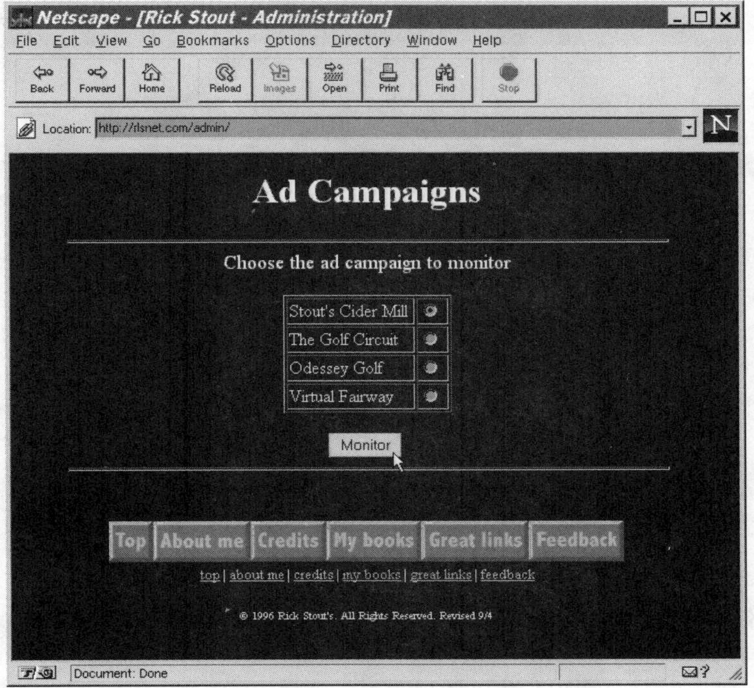

Figure 14-4. *The intro page for monitoring ad banner activity*

I named this CGI script adtrack.pl. It produces the statistics page shown in Figure 14-5. As I mentioned, the first time you call this script (from the Ad Campaigns page), you pass in the name of the campaign through the form's field. If you're like me, you might decide that you'd like to leave this page visible on your computer's desktop as you work and have the statistics update automatically. I like them to update every minute—this way, it's not so frequent as to distract me, but I can still say that my ad statistics are up to the minute.

You can make a web page refresh itself periodically by using the HTML <meta> tag with the refresh option. This is easy enough to do, but in this case, there's a wrinkle. There are two pieces of information you have to supply in the meta tag: the frequency (in number of seconds) that page should refresh

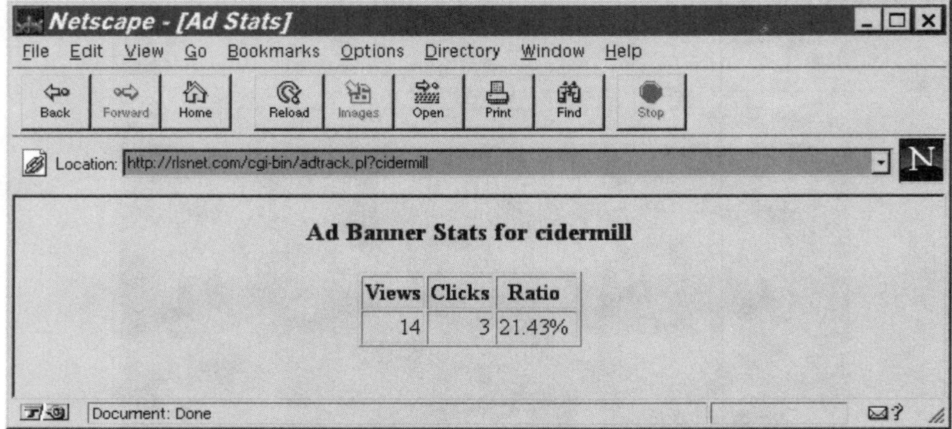

Figure 14-5. *Monitoring banner activity for the cidermill campaign*

and the URL. In this case, the URL is the same—you just want to rerun the CGI program again. But this time, you're not calling it from a web page form. How do you get the value of the field (cidermill) into the program so that it knows which campaign to provide statistics for?

I chose to make the CGI script accept a parameter in addition to the form field through standard input. When the script runs, it first checks to see if a parameter was passed in. If not, it can assume that there will be a value to use for the filename coming from the form. Otherwise, it uses the parameter that was passed in for the log file name. The complete source for adtrack.pl is listed in Figure 14-6.

The adtrack.pl script checks to see that it has both a views and clicks log file for the given campaign, then it counts the lines in each log file and presents the results in an HTML table. The final piece is to calculate the click-through ratio, which is simply the number of clicks divided by the number of views.

An alternative to the way I've chosen to monitor each ad or ad campaign separately would be to monitor them all on a single page instead of choosing one particular ad or campaign to watch. This way, you can see the status of all your ad banners at the same time. If you're getting paid by the impression or the click-through, you can even add a column showing the dollars rolling in!

```
#! /usr/local/bin/perl
# ADTRACK - Summarizes Ad Banner Activity
#
# Call from an HTML form:
# <form method=post action="/cgi-bin/adtrack.pl">
# with a single field named "campaign" or an argument
# representing the name log name
require "cgi-lib.pl";

$docroot = '/var/www/docs/rlsnet';

if ($ARGV[0]) {
  $name = $ARGV[0];          # log name is passed in by refresh
} else {
  &ReadParse(*FORM);         # log name from form field
  $name = $FORM{'campaign'};}

$viewlog  = "$docroot/adlogs/$name"."_views.log";
$clicklog = "$docroot/adlogs/$name"."_clicks.log";

print <<EOT;
Content-type: text/html\n\n
<html><head><title>Ad Stats</title>
<meta http-equiv=\"refresh\" content=\"5;
url=/cgi-bin/adtrack.pl?$name\">
</head><body><center><h3>Ad Banner Stats for $name</h3>
EOT

if (!((-f "$viewlog") && (-f "$clicklog"))) {
  print "<h3>Missing a log file!</h3>";
  finish();}

if (!(open (VLOG, "$viewlog"))) {
  print "Can't open view log for $name!";
  finish();
}

if (!(open (CLOG, "$clicklog"))) {
  print "Can't open click log for $name!";
  finish();
}

while (<VLOG>) { $views++; }
while (<CLOG>) { $clicks++; }
$ratio = sprintf("%.2f\%", ($clicks/$views)*100);

print "<table border><tr><th>Views<th>Clicks<th>Ratio
       <tr><td align=right>$views<td align=right>$clicks<td align=right>$ratio
       </table>";

sub finish
{
  if (VLOG) {close VLOG};
  if (CLOG) {close CLOG};
  print "</body></html>";
  exit;
}
```

Figure 14-6. *The complete adtrack.pl program, which displays rudimentary statistics for ad activity and automatically updates its display every minute*

One final note of qualification: This script uses a very simplistic algorithm to determine the number of views and clicks—it just counts the number of lines in the log file. In practice, you might want to replace this code with something more sophisticated, perhaps by keeping track of the last time stamp each time you read the file. You might even want to consider showing impressions and click-throughs by the day or by the hour.

Selecting Ads from a Play List with CGI

Progressing in complexity from the simple, single static banner ad, next we'll discuss serving ads dynamically from a list. Such a list is sometimes called a *play list*, perhaps in reference to the lists that radio station disc jockeys use for playing songs on the radio.

There are several good reasons why you might prefer to serve up a collection of ads instead of always having the same ads on every page. For one, if your ads never change, your readers may get used to them and cease to even see them anymore; if you're constantly swapping ads, readers are more likely to notice every one of them.

Another reason to try a number of ads is to evaluate the effectiveness of each of them. I'm by no means an artist—graphic or otherwise—and I'm continually amazed at the difference I find in the effectiveness of different ad banners. There's some magic in designing and creating an effective ad; I can't tell you what it is or how it works, but I can tell you how to determine if an ad is working.

By serving a collection of ads for the same resource or site and logging the impressions and click-throughs for each, you can see some amazing results. Some ads will generate much higher click-through ratios than others. By keeping track of them, you can weed out the ineffective ads and replace them with new ones.

As you might have guessed, I've got another Perl script for you in this section. This script (I've called it playlist.pl) is based on the earlier logview.pl, since it does essentially the same thing—it displays a banner ad and logs the impression. Following are the major differences between logview.pl and playlist.pl:

- Playlist.pl takes only one argument—the name of the campaign from which the script will construct filenames. We don't need the second argument from logview (the name of the banner graphic to display), because we're going to be selecting banner filenames sequentially from a list.
- Playlist.pl uses a file in which the filenames for each graphic are stored, one to a line. I keep these files in my /admin directory. The play list for each campaign has the format <*name*>.playlist, where <*name*> is the name of the campaign.
- Playlist.pl uses another file in the /admin directory with a name in the format <*name*>.lastad. This file stores a number representing which ad was last served (with 1 being the first ad in the list, 2 the second, and so on).
- Playlist.pl logs the filename of the banner served in the view log in addition to the time stamp and IP address.

Note that there is no attempt here to serve these ads randomly. Instead, I've consciously chosen to serve them sequentially. Although every reader won't necessarily ever see all of the ads, this will ensure the most even exposure overall.

Figure 14-7 presents the full source for playlist.pl. After processing the argument, setting up the variables, and computing the time stamp, the script opens the lastad file to read the line number of the last-served ad. It increments this number and writes it back out to that file. The incremented number gets stored in the $this variable, which represents the line number of the ad the script is going to serve in the current run.

Next, the script reads $this lines from the playlist file to get the name of the graphic file. Finally, it logs the view event and outputs the graphic file to the reader's browser.

Of course, if you're actually going to use this script to serve ad banners dynamically and the script in the previous section (adtrack.pl) to evaluate each of them, you will need to use separate log files for each. But this modification shall be left as an exercise for the student.

```
#!/usr/local/bin/perl
# PLAYLIST - Serves and Logs banner views from a list
#
# Called from an HTML link of the form:
# <a href=/cgi-bin/playlist.pl?LOGNAME>

$docroot = '/var/www/docs/rlsnet';
$name    = $ARGV[0];        # name of the campaign
$logfile = "$docroot/adlogs/$name"."_views.log";
$adlist  = "$docroot/admin/$name".".playlist";
$lastad  = "$docroot/admin/$name".".lastad";

($sec,$min,$hour,$mday,$mon,$year,$wday,$yday,$isdst) =
    localtime(time);
$tstmp = sprintf("%02d/%02d/%02d:%02d:%02d:%02d",
    $year,$mon,$mday,$hour,$min,$sec);
if (!(-f "$logfile")) {          # create log if needed
  open (FILE, ">$logfile");
  close FILE;
}

if (!(-f "$lastad")) {           # create lastad file if needed
  open (FILE, ">$lastad");
  print FILE "0";
  close FILE;
}

if (open (FILE, "$adlist")) {    # count the number of ads
  while (<FILE>) {
    $adcount++;
  }
  close FILE;
}

if (open (FILE, "$lastad")) {    # increment the ad number
  $last = <FILE>;
  $this = $last+1;
  if ($this > $adcount) {
    $this=1;
  }
  open(FILE, ">$lastad");  # save the new ad number
  print FILE sprintf("%d", $this);
  close FILE;
}

if (open (FILE, "$adlist")) {   # get the file name from the list
  for ($i=0; $i<$this; $i++) {
    $adname = <FILE>;
  }
  close FILE;
  chop($adname);
  $picfile = "$docroot/ads/$adname";
}
if (open (FILE, ">>$logfile")) {# log the event
  print FILE $tstmp, ," ", $ENV{'REMOTE_ADDR'}, " ", $adname, "\n";
  close FILE;
}

if (open(FILE, "$picfile")) {   # output the graphic
  print "Content-type: image/gif\n\n";
  while (read(FILE, $foo, 1024)) {
    print $foo;
  }
  close FILE;
}
```

Figure 14-7. *The playlist.pl script serves ads*

Rotating Banners

In the previous section, you learned about a way to serve ads dynamically from a play list with a Perl CGI script. This is considerably more professional and effective than simply having static ad banners on your pages. However, sometimes you want even more flash; especially on pages that have a longer average view time (lots of stuff to read and look at), you may want animated ads or ads that change by themselves. This is a real attention grabber. Such an ad is called a *rotating banner.*

Besides the motivation of serving a real eye-catching ad, there's another reason you may want to consider rotating ads. If you get paid by the impression, you can serve an impression of multiple ads while someone is idly reading or watching your page.

Right now, there are two ways to accomplish this: with animated GIF images and Java applets. Neither way is ideal, but for now, it's all we have.

Animated GIFs

Achieving rotating banners with animated GIF files is the less desirable way to rotate ads. GIF animation was really designed for animation and not for storing multiple images in a single graphic file. But that is how you can use an animated GIF to display a rotating ad.

The tools to build animated GIFs are freely available on the Internet. Go to any of the major search engines and do a search on "GIF construction set." This is the bundle of tools and utilities you will need.

A classic example of an application for an animated GIF is a banner ad with a small part of it that is animated—say, a rotating globe or logo. With the tools that you use to construct animated GIFs, you save the initial banner image, then rotate your globe or logo a small amount, and save the next image. You proceed in this manner until you've shot a picture of your graphic in each step of its animation. Then you turn loose the software that produces the finished file, and it does its magic.

When readers load a page with an animated GIF in their browser, it proceeds to do its animation thing. If you craft the animated GIF in such a way that it loops continually, it will continue its animation until the reader moves on to another page.

Although animated GIFs normally have a relatively high number (15 or 20 isn't uncommon) of different embedded images, where changes between the images are only slight, you can have a much smaller number (like two or three images) that are more drastically different.

There are several major drawbacks to using animated GIFs for ad rotation. For one, the size of an animated GIF file can grow quickly. A file with two embedded graphics that are completely different will usually be more than the combined size of both. Once the graphic is loaded in the reader's browser, this isn't an issue, but getting your readers to wait for it to load the first time might be.

The second major drawback is there is no sure way to tell if a reader actually sees a second or third image. Even if you went to the trouble of checking the bytes transferred for the image (to make sure the entire file transferred) and the length of time readers viewed the page (to see if they viewed it long enough for the animated GIF to roll over to the next image), it would still be a toss-up whether readers actually got to see images after the first one.

Finally, since the magic of an animated GIF is in the graphic file itself, there's no way to provide more than a single click-through target for the graphic. This pretty much limits animated GIFs to being dedicated to a single company or product. So you might as well just consider it a single, animated ad rather than a rotated ad.

Using Java to Rotate Ads

The preferred way to rotate ads is with Java—and you don't necessarily have to be able to write Java code to use it. There are hundreds (if not thousands) of Java applets freely available on archive sites all over the world. In addition to these, numerous companies (like Macromedia, at www.macromedia.com) allow you to download professional-quality Java applets freely.

One of the applets that Macromedia distributes is called Banners.class. This is the premier general-purpose banner applet available today. This applet is fully configurable with its configuration file and the HTML from which you call it. All you have to do is create the HTML code to call it, create the configuration file, and have the Banners.class applet in your document tree.

You can define multiple banner ads for the applet to cycle through—each with a different background image and text that floats over the images. In addition, you can define parameters like the entry and exit styles (fade in/out, wipe up/down, etc.), shadow colors and depth, text fonts and colors, text scrolling and traversing, and separate URLs to associate with each banner.

The HTML you write to embed a Java applet on a page is straightforward. Here's the code that implements the Banners.class applet on the golf site:

```
<applet code="Banners.class" width=500 height=65>
<param name=bgColor    value=black>
<param name=bgEnter    value=squeeze>
<param name=bgExit     value=squeeze>
<param name=fps        value=50>
<param name=cpf        value=1>
<param name=pause      value=3>
<param name=msgFile    value="index.txt">
<param name=reloadInterval=180>
 <hr>
   If you were using a Java-enabled browser,
   you would see an animated scrolling text sign
  <hr>
</applet><P>
```

You reference a Java applet in your HTML with the <applet> tag. The additional parameters set up defaults that apply to the banner as a whole (as opposed to the different banners that the applet will display).

Notice that the value for the msgFile parameter is index.txt. This is the configuration file where you describe the different banners the applet is to display. (You can have any filename you like; we just happened to choose index.txt.) Figure 14-8 shows the contents of our index.txt file.

Although we haven't used it with our implementation of the Banners.class applet, the {url=} parameter is one of the additional parameters you can include. This is how you define a different URL (click target) for each banner. We don't include that parameter here, because we're not using the applet to rotate ads—we're using it to display the results of recent golf tournaments throughout the world, as shown in Figure 14-9.

Although you've probably figured this out for yourself, I'll just give you a tip to help connect the dots. Since you can define separate URLs for each banner, those URLs can be separate CGI scripts (like the logclick.pl script we looked at earlier in the chapter) to log the click-throughs for each different banner the applet displays.

In this chapter, you've learned about banner ads—both simple ones and more complex ones that you can use to log impressions and click-throughs.

```
- The BC OPEN -- FINAL RESULTS
\n1. Fred Funk              -16          F
\n2. Peter Jordan           -16          F
\n3. Patrick Burke          -13          F
      {style=bold}
      {bgImage=pgaback.jpg}
      {font=courier}
      {size=12}
      {shadowColor=black}
      {shadowDepth=2}
      {textColor=yellow}
      {enter=wipeDown}
      {exit=wipeUp}
      {align=center}
      {pause=5}
|-Solheim Cup-Final Rounds Complete
\n United States 17  -  Europe 11
      {style=bold}
      {bgImage=lpgaback.jpg}
      {font=courier}
      {size=18}
      {shadowColor=black}
      {shadowDepth=2}
      {textColor=yellow}
      {enter=wipeDown}
      {exit=wipeUp}
      {align=center}
      {pause=5}
```

Figure 14-8. *The configuration file for the Banners.class applet holds the information about each of the banners that the applet will rotate, including the background image, the foreground text, entry and exit styles, colors, and so on*

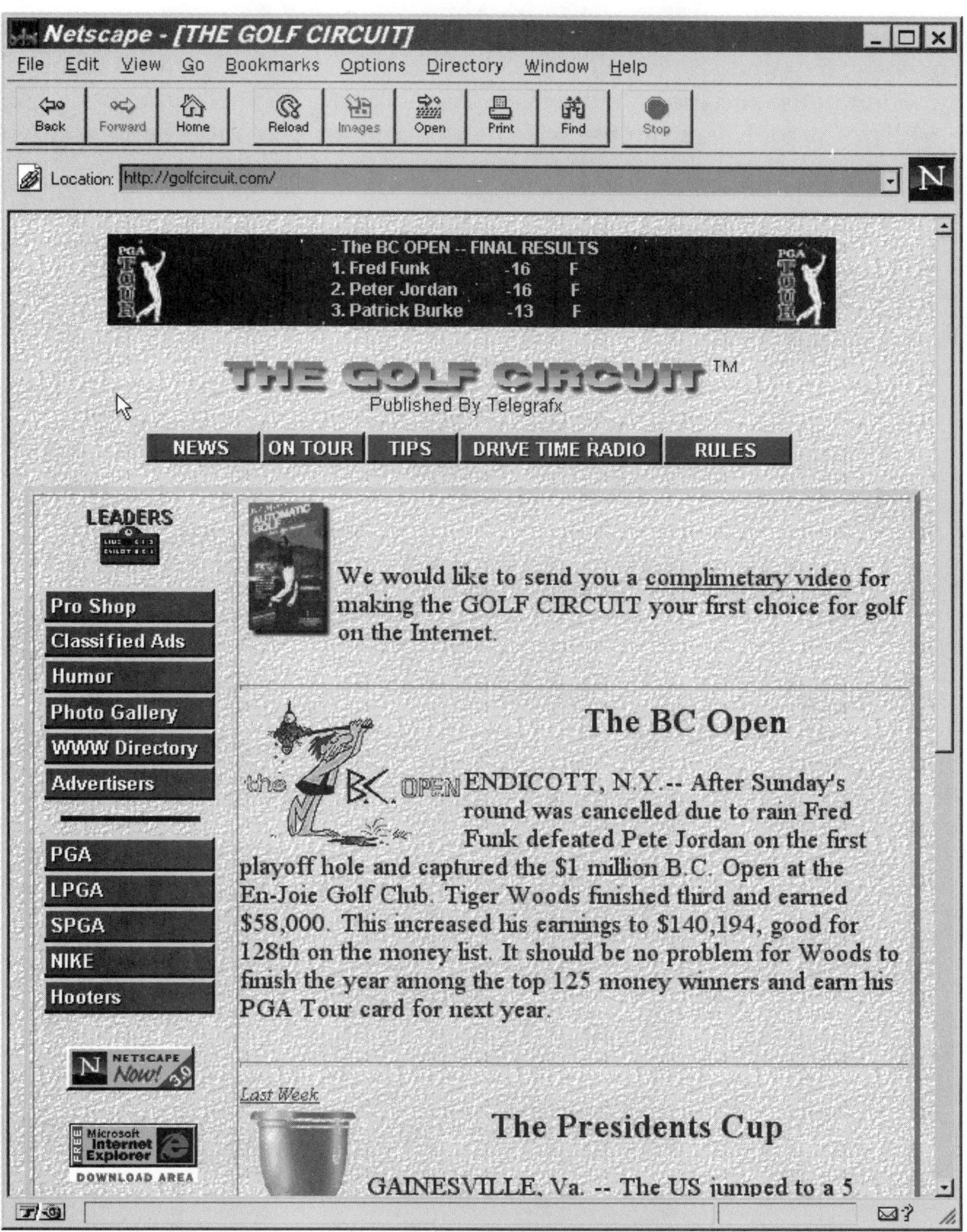

Figure 14-9. *The Banners.class applet running on the golf site (it's displaying the banner at the top of the page)*

You've also learned about serving ad banners from a play list and rotating them with animated graphics files and Java. In the next chapter, we'll take it up another notch and look at ad servers and ad networks and some of the really nifty tools you can use with them to mount even more serious ad campaigns.

Chapter Fifteen

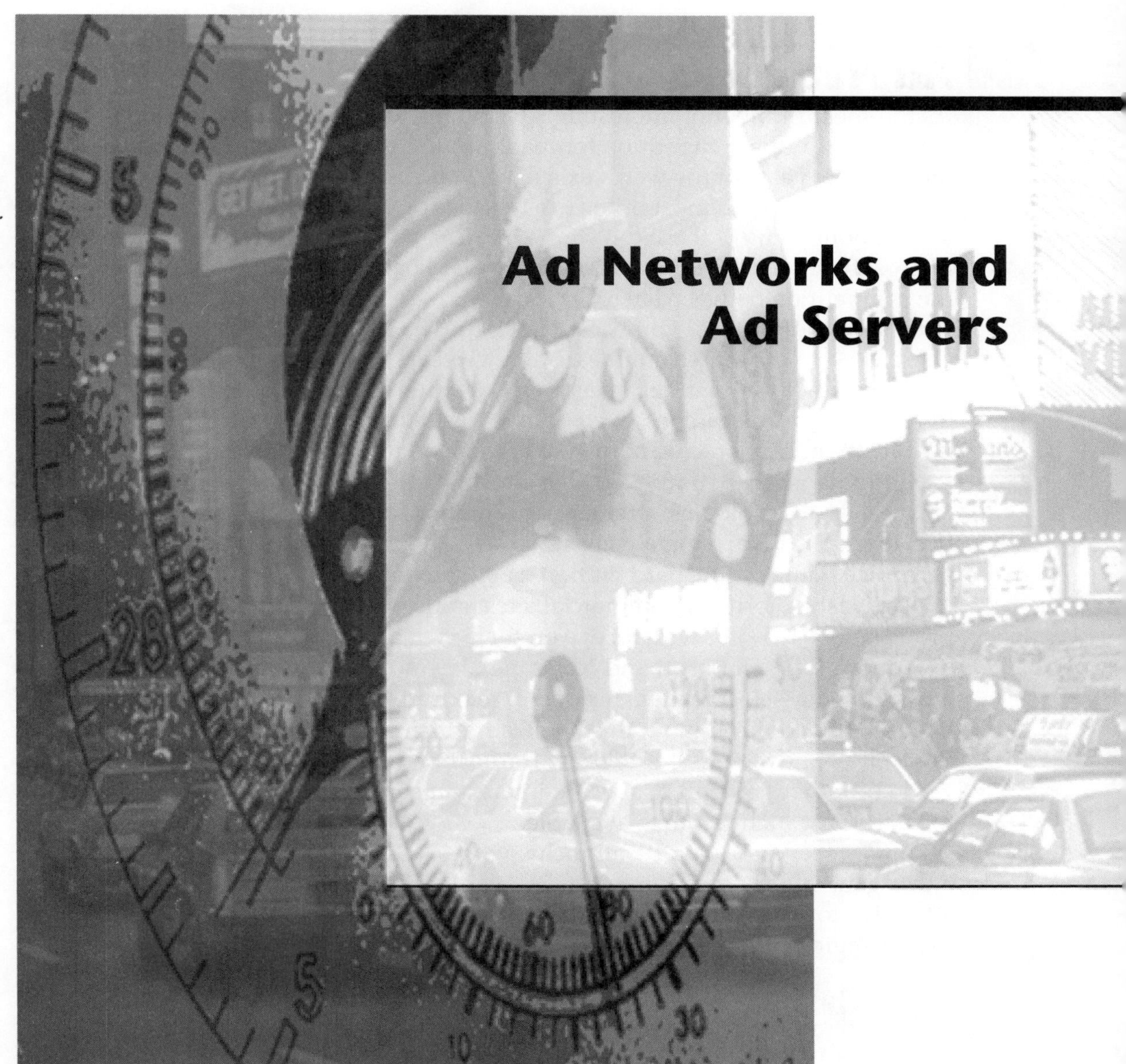

Ad Networks and Ad Servers

In the last chapter, you learned some simple and inexpensive techniques for logging ad views and click-throughs using a small collection of Perl scripts. In this chapter, we're going to explore ad networks and commercial ad servers—the professional upgrade to the "poor webmaster's" ad tracking system we covered in Chapter 14.

Ad Networks

As the sheer number of interesting sites on the web continues to explode exponentially, the competition for reader traffic is getting hot. For much of the first few years of the web's existence, any experienced web surfer could probably list for you most of the top ten sites on the web. Now, we all still know about a few of these top sites—like Yahoo, Infoseek, and maybe the Microsoft network. At the next level, the field has become so populated and the competition between sites for traffic so fierce as to resemble a battleground.

It used to be that all you had to do to draw a lot of traffic to your site was get it listed in the search engines. But after years of accepting hundreds—even thousands—of new listings every day, the search engines have so many entries in them that most searches, even on the most arcane of keywords, yield more links than most people want to take the time to follow and explore.

Building a highly trafficked web site today involves a whole new set of techniques, technologies, strategies, promotion, and a whole lot of hard work. If your goal is to develop a web site to sell advertising space, then you've got no choice but to do the hard work it's going to take.

The vast majority of commercial sites on the web are company or product-focused sites with an eye on their own promotion. With very few exceptions (like Microsoft and Netscape), it's unlikely such a site will ever attract the masses of readers needed to make advertising profitable. But, as I've mentioned throughout this book, advertising isn't usually the goal of these sites. Site owners out to draw traffic for advertising know that they've got to be either general- or specific-interest in nature and they've got to strive to be the best such resource available.

It may seem unnaturally all-inclusive to say that these sites have to be either general- or specific-interest. By general-interest, I mean a site has to be attractive to a broad cross section of the web-browsing public; for example, search engines are used by virtually everyone, and many people visit news-

and politics-related sites. On the other hand, special-interest sites (which are sometimes referred to as *affinity* sites) cater to the specific tastes of a smaller population but enjoy a higher penetration rate.

The golf site we've used as an example in this book is such a site. There are far fewer people genuinely interested in golf than there are people interested in world and national news, but golf fans actively seek out golf-related resources on the web. When they find our site, they bookmark it and visit often, because we do the hard work it takes to keep it current with news, events, and new material. The net result is that we garner a higher percentage of the golf-interested web browsing population than a news site would of readers interested in headline news.

The point I'm trying to make here is that there are very few instant successes. To build a highly-trafficked site requires many months of hard work directly focused on making the most compelling site in the world. Measuring your traffic and tailoring your site, as we've discussed in this book, is just part of it. You also have to spend countless hours getting yourself situated optimally in the search engines, building online relationships, and swapping links with other related sites of similar interest. You have to do all this in addition to staying on top of the content (news, current events, and so on) in your site—which must always remain the highest priority.

Many folks have made the mistake of thinking that if they create a web site and just get it listed in the search engines, then floods of eager readers will come browsing. It just doesn't work that way. It takes literally months of hard work focusing on content and building a reputation to draw enough visitors to warrant any ad revenue at all.

Sometimes, after all of the hard work, it still isn't enough. Many site owners toil away for months crafting a web site and find that they still lack the necessary traffic to warrant a serious look by a major advertiser. This is where ad networks come into the picture.

An ad network is an association of several (or even many) web sites that pool their traffic to together deliver the ad impressions that none of them can do alone. An ad network is usually served by a single, central ad server that serves ad banners and logs both impressions and click-throughs along with the usual reader demographics reported by common log analysis packages (domains, browser type, operating system, etc.). Although we've referred to the collection of scripts in Chapter 14 as a "poor webmaster's" ad tracking system, you could really use it to serve ads from a central site as well. We'll look more at this in the next section, along with a good example of first-class ad server software.

The largest ad network today is the DoubleClick network (at www.doubleclick.com). The major benefit of an ad network is that it handles the details of selling the ad space made available by the members. This is a big benefit to most small- and medium-sized sites that don't want, or can't afford, to employ their own sales staff.

The downside of the big ad networks like DoubleClick is that while they accept membership applications from virtually anyone, you do have to be able to deliver a minimum of 100,000 impressions per month for them to seriously consider placing an ad on your site. That's not total page views—it's views on pages on which you would want to display ads.

Even this amount of traffic can be a problem for many medium-sized sites. Remember back in Chapter 9, where we ran the Golf Circuit log files through the commercial analysis packages; the top ten pages on the site drew a total of around 53,000 views—only about half of what it takes to be considered by DoubleClick. And we (Tom, the site's owner, and I) considered the Golf Circuit a bustling site at 27,000 visits per month!

So, while the big ad networks like DoubleClick do deliver on their promise to sell and place advertisements, it's clear that they still prefer to be doing business with sites we can only classify as high volume. What do you do if your site doesn't have enough traffic to warrant a second look from the big ad networks like DoubleClick? This is a tough problem that plagues many owners of small- and medium-sized web sites, but there are solutions.

One is to form your own ad network. There are a couple of ways you can do this. At the low end, you can make arrangements with other, similar sites on the web (like your competitors) who also probably aren't making any money yet. By forming a loose coalition or partnership and using even rudimentary ad server software like the scripts we developed in Chapter 14, you may be able to coordinate enough impressions to collectively go to the big networks.

Another way is to go with a completely new and different kind of ad network, such as the Real Media Network. Real Media is just as interested in small sites as they are in large. In fact, the core of their business model encourages the diversity that many smaller sites have to offer. Real Media wants at least medium-sized sites that have a potential to deliver impressions, but they make the transition from small site to large much easier.

For participating in the Real Media Network, Real Media provides their excellent ad server software, Open AdStream, free of charge. We'll take a detailed look at Open AdStream in the next section. This software is state of the art for managing and reporting on sites, ad networks, and ad campaigns.

Ad Servers

At its simplest, an ad server is simply web server software running on a computer connected to the Internet that serves ad banners when they're requested. In Chapter 14, we went through the exercise of creating a Perl CGI script that logs a request for a banner ad, then shoots the ad out to the requesting reader's browser. In our example, the CGI program was on the same computer as the HTML pages we were serving, but that doesn't have to be the case.

Recall that when we finished both Perl scripts (the one that logs views and the one that logs click-throughs), the HTML code with which we linked to our rudimentary ad server was the following:

```
<a href="/cgi-bin/logclick.pl?cidermill+http://cidermill.com">
<img src="/cgi-bin/logview.pl?cidermill+/art/banner3.gif">
</a>
```

Notice that the href parameter for the anchor tag and the src parameter for the image tags just reference a URL. In this case, the URL was relative—referring to other files in the same document tree and on the same server as the file containing this HTML.

However, there's no reason at all why these scripts must be on the same system. Instead, they could be in the web server's document tree of a completely different server running on a completely separate computer down the hall or halfway around the world. For example, the tags could look like this:

```
<a href="http://ad.server.com/cgi-bin/logclick.pl?cidermill+http://cidermill.com">
<img src=" http://ad.server.com /cgi-bin/logview.pl?cidermill+/art/banner3.gif">
</a>
```

Provided that the banner ad graphic is on that system, which is otherwise set up to handle it, this scenario will work just as well as serving ads from a single system.

Realizing this, people have begun to think about the merits of serving and logging ad activity centrally, and some (many, in fact) entrepreneurs have jumped in, developing businesses based on this idea. Some industry analysts have even gone so far as to estimate one web-based advertising company now for every seven commercial web sites. This doesn't make it sound like there's still much of an opportunity here—especially when these companies are based on technology so simple that you know much of it if you've read Chapter 14. But the companies that will be the survivors among these advertising companies are continually coming up with new innovations, improved reporting, and enhanced capabilities that are far beyond what we did in Chapter 14.

Your exposure to ad servers in this chapter is going to consist mainly of a tour through a real commercial ad server. We chose Real Media's ad server software, Open AdStream, as a model ad server system to demonstrate the level of sophistication possible with commercial ad server software—or as Real Media calls it, an *advertising planning and placement software engine.*

Real Media's Open AdStream

Real Media's flagship product—service, actually—is a network that bands together local online newspapers and other publishers to make them attractive media for national and regional advertisers. National advertisers find this beneficial because of the strengths of local publishers' local content, trusted brand identity, and the established relationships that these local publishers have with the consumers and businesses in their markets. In essence, this is an ad network. To facilitate the use of their network, Real Media developed the Open AdStream software.

Like our few Perl scripts in Chapter 14, the Open AdStream software is a collection of CGI programs. Real Media has woven them together tightly and integrated them with a sophisticated database management system to produce a highly effective set of tools for managing and implementing ad campaigns.

Your first glimpse of the Open AdStream software is the main menu. Shown in Figure 15-1, it has hypertext links for the different modules of the server:

- Advertisers, for maintaining a list of your advertisers
- Users, for adding and maintaining accounts for people who have permission to work on your ad campaigns or see reports

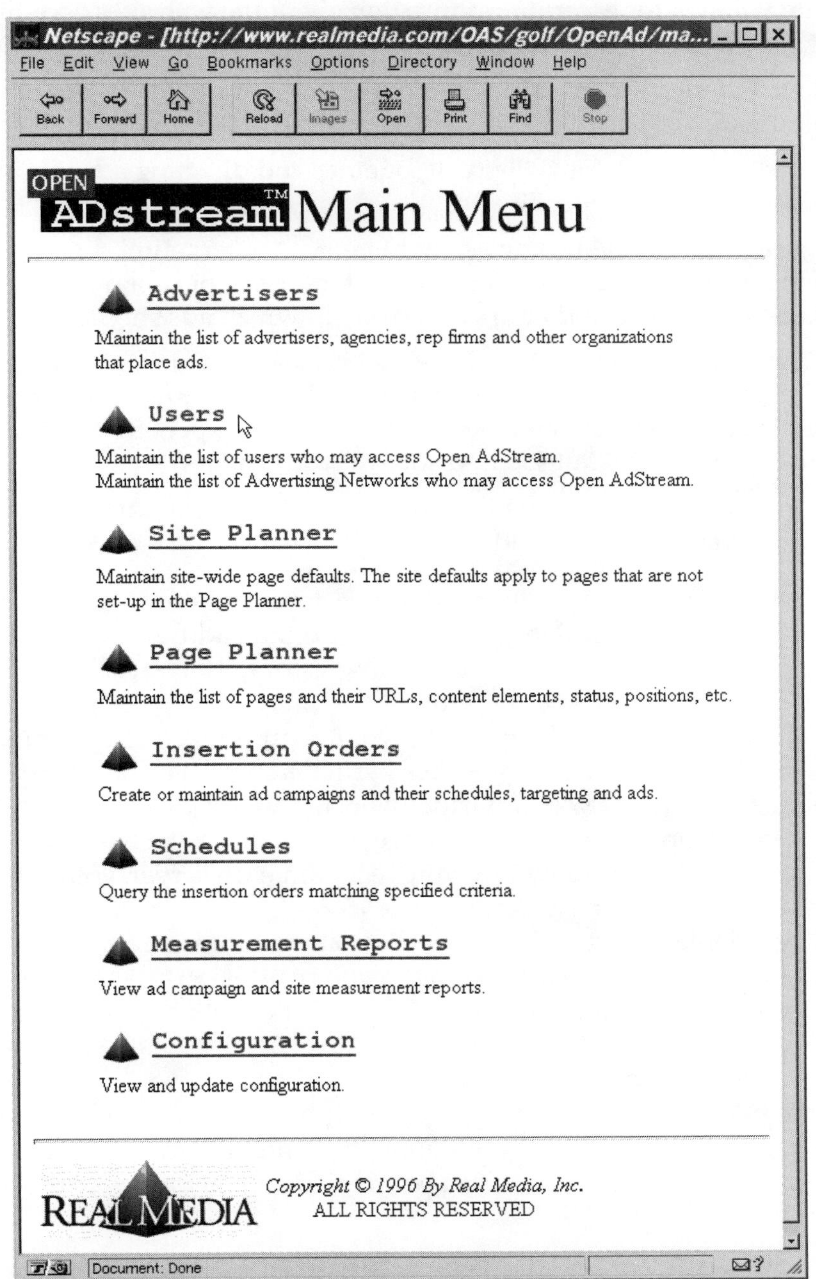

Figure 15-1. *Real Media's Open AdStream main menu*

- Site Planner, for entering information about the web sites on which you will be managing ad campaigns
- Page Planner, for specifying the pages on each site that are candidates to run ads
- Insertion Orders, for tying it all together and directing ads to run on specific pages on specific sites
- Schedules, for fine-tuning ad placement, as in targeting specific states, regions, browser users, operating system users, or root domains and the days of the week and hours of the day in which ads should run
- Measurement Reports, for checking on the status of ads, sites, and campaigns, including the impressions, click-throughs, ratios, and all of the other normal demographics you would expect from any log analysis package (but focused solely on ad readers)
- Configuration, for setting up or maintaining general information, like your e-mail addresses and so on

The first step in using the Open AdStream software is to set up your advertisers and administrative and supervisor e-mail addresses and add your authorized user accounts to the system. This is straightforward. You just fill in the forms and click on the submit buttons.

Next, you tackle the Site Planner, shown in Figure 15-2. Here's where you enter information about each of the sites participating in your ad campaign. If you manage several servers or virtually hosted sites, by setting them up separately in the Open AdStream software you can track the effectiveness of each of the servers separately. Once you've added each server, your Sites page will have a table entry for each (see Figure 15-3).

The Page Planner module is where you maintain information about the specific pages on each site participating in your campaign on which you plan to display ad banners. Each page has its own ID, the path and filename, and a description. In Figure 15-4, you can see the Add Page link just above the table of pages defined to run ads.

Under the Add Page link on this page is where you specify pages to run ads. For each page to run an ad, you can define the content of the page and the ad positioning on the page so that ads can be targeted more narrowly to the reader.

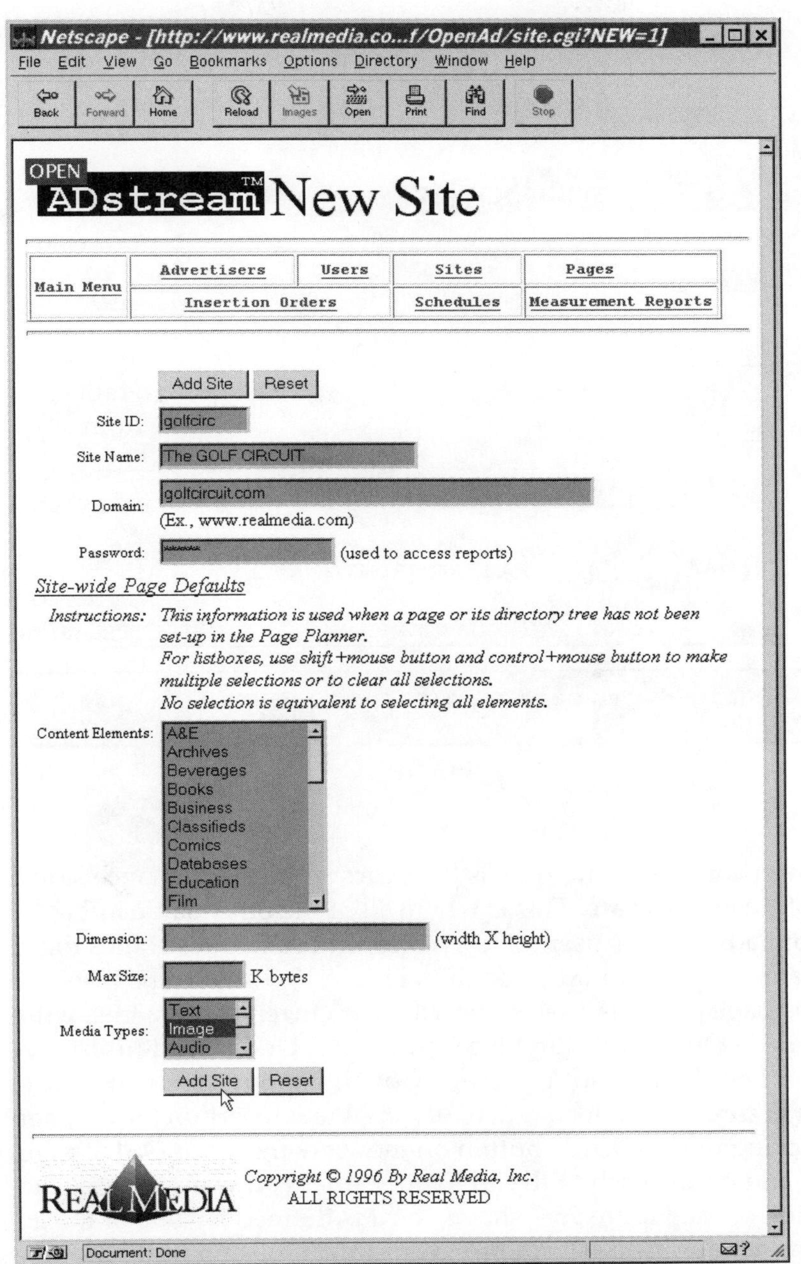

Figure 15-2. *AdStream's New Site form for adding a new site under your campaign control*

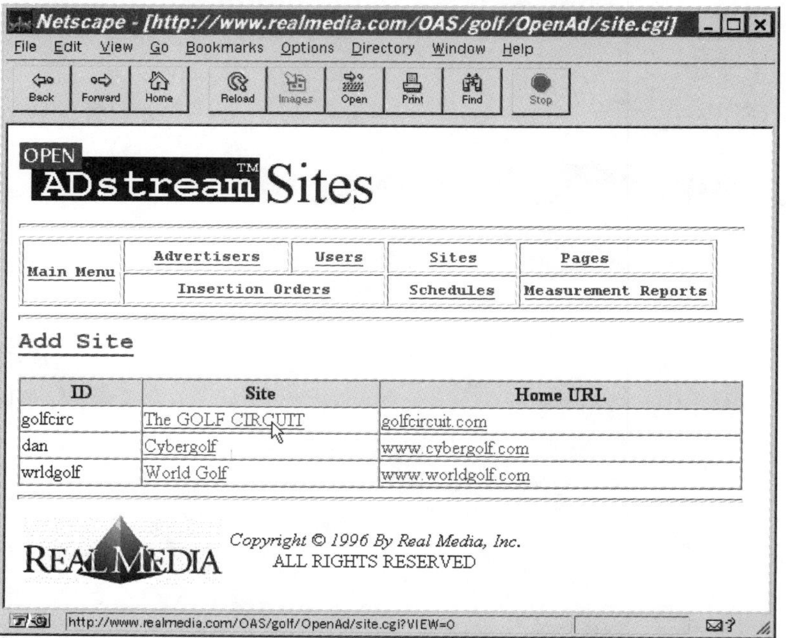

Figure 15-3. *After you add a site, it shows up in the Sites table*

The Insertion Orders module is the heart of the maintenance side of the Open AdStream software. This is where all of the other static information you enter about advertisers, users, sites, pages, and schedules comes together and you actually define your ad campaign. Figure 15-5 shows the Insertion Orders page. This page presents a table showing the current campaigns, with links to Add Insertion Order and Run Live Campaigns. Under Add Insertion Order is where you first define your campaigns, by filling out a form like the one shown in Figure 15-6 (which is actually the View Insertion Order page).

By clicking the View Ads button on the View Insertion Order page, you can view and define each of the ad banners that you want to associate with an insertion order. Figure 15-7 shows the ads defined for one of the golf site campaigns. Under the New Ad link, you can define and upload a new banner.

The Measurement Reports module is your entry point to the reporting facilities of the Open AdStream software. The initial link takes you to a Reports page, where you specify the date range for which to view reports and optional fields to narrow your reports to a particular site or campaign. Clicking the List Reports button displays the available reports under your criteria (see Figure 15-8). Click the Report link to access the reports.

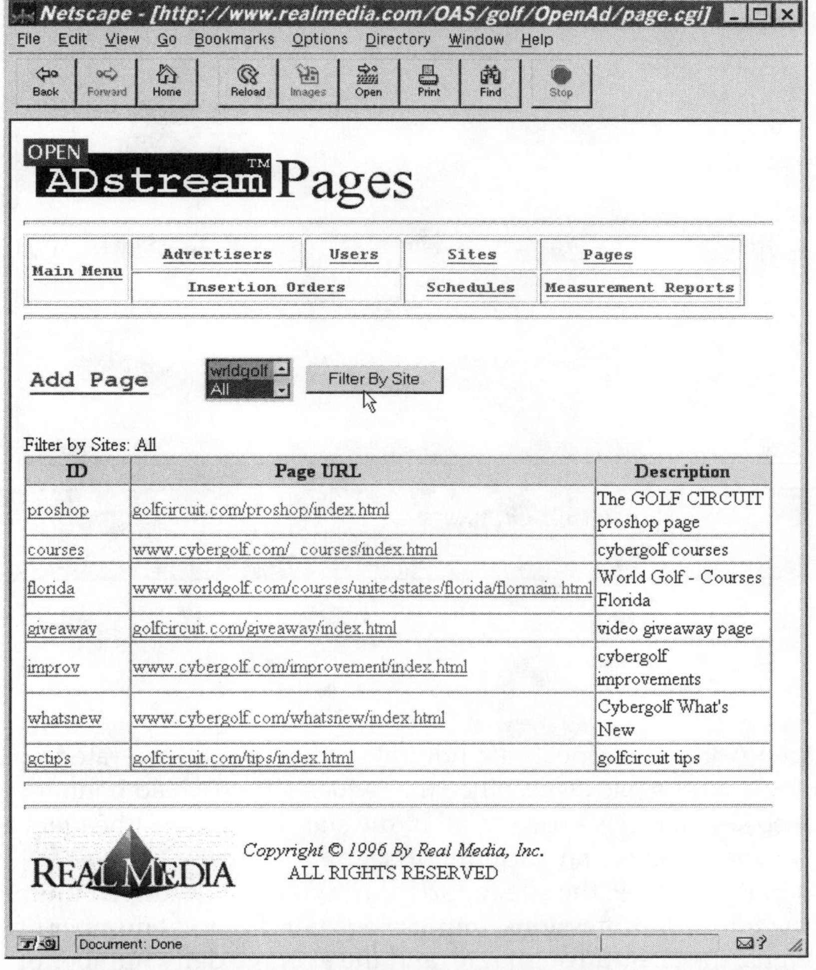

Figure 15-4. *Open AdStream's Page Planner shows each of the pages you've defined to run an ad banner*

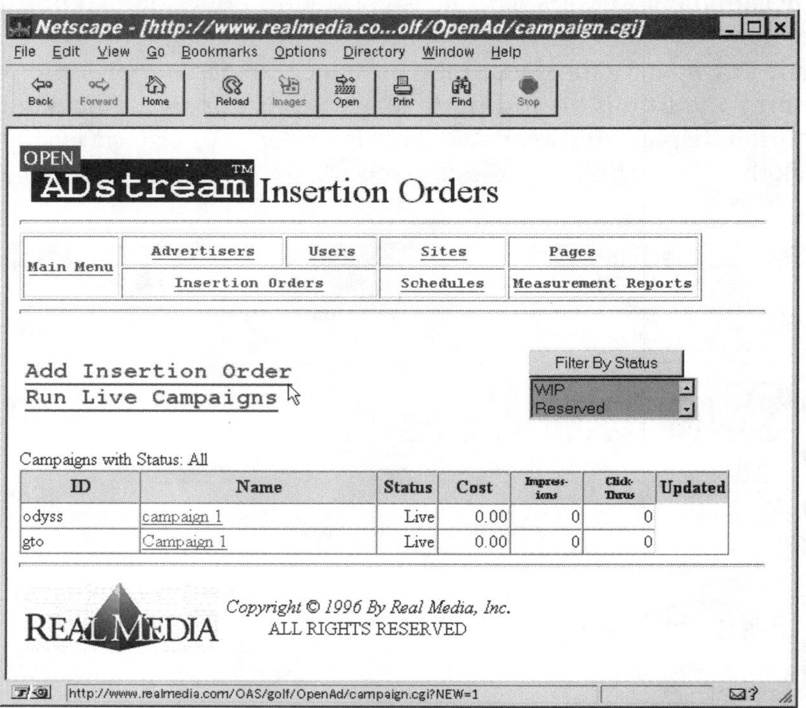

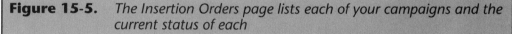

Figure 15-5. *The Insertion Orders page lists each of your campaigns and the current status of each*

The Open AdStream reports are not unlike the reports generated by the high-end log analysis packages. Since the requests for each ad banner originate from the web site reader's browser, all of the standard request headers are available for inclusion in statistical reports.

Shown in Figure 15-9, the site reports start with an Executive Summary, showing the total ad impressions, impressions per day, total number of click-throughs, the click-through rate, and the average daily number of

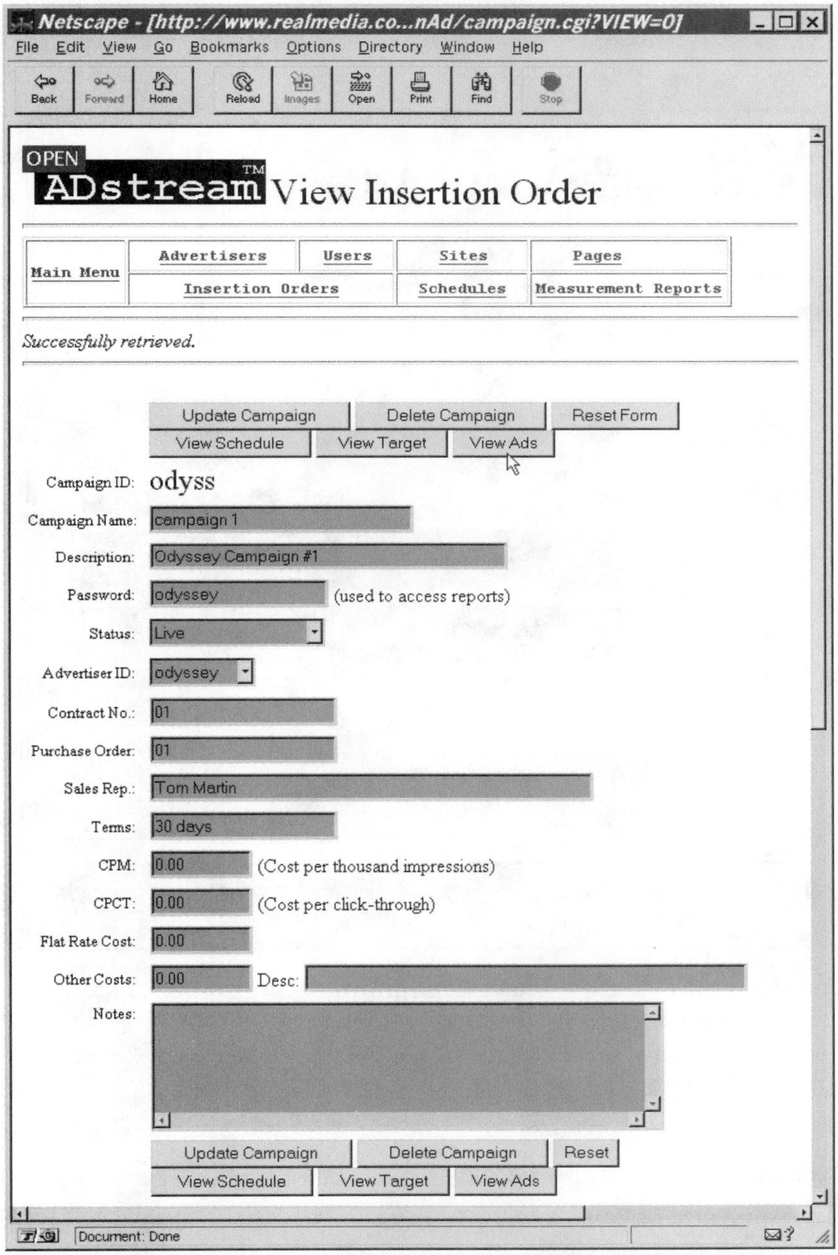

Netscape - [http://www.realmedia.co...nAd/campaign.cgi?VIEW=0]

File Edit View Go Bookmarks Options Directory Window Help

Back Forward Home Reload Images Open Print Find Stop

OPEN ADstream™ View Insertion Order

Main Menu | Advertisers | Users | Sites | Pages | Insertion Orders | Schedules | Measurement Reports

Successfully retrieved.

Update Campaign | Delete Campaign | Reset Form | View Schedule | View Target | View Ads

Campaign ID: odyss

Campaign Name: campaign 1

Description: Odyssey Campaign #1

Password: odyssey (used to access reports)

Status: Live

Advertiser ID: odyssey

Contract No.: 01

Purchase Order: 01

Sales Rep.: Tom Martin

Terms: 30 days

CPM: 0.00 (Cost per thousand impressions)

CPCT: 0.00 (Cost per click-through)

Flat Rate Cost: 0.00

Other Costs: 0.00 Desc:

Notes:

Update Campaign | Delete Campaign | Reset | View Schedule | View Target | View Ads

Document: Done

Figure 15-6. *With the New (not shown) and View Insertion Order pages, you set up and maintain your ad campaigns*

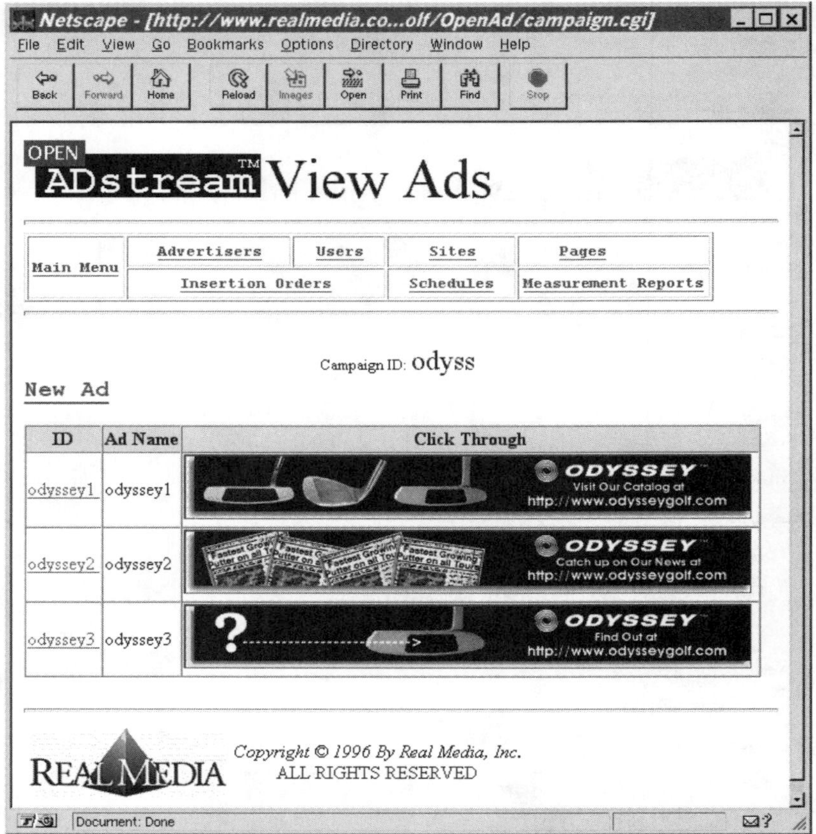

Figure 15-7. *The View Ads page shows each of the ad banners which have current insertion orders*

click-throughs. Following the Executive Summary are statistics tables and graphics on the following:

- Top ten states by impressions
- URLs by impressions
- Ad campaigns by impressions
- Ad images by impressions
- Impressions by time of the day
- Impressions by day of the week
- Impressions by country

- Impressions by top-level domain
- Top ten domains of visitors
- Impressions by browser type
- Impressions by Netscape version
- Impressions by operating system

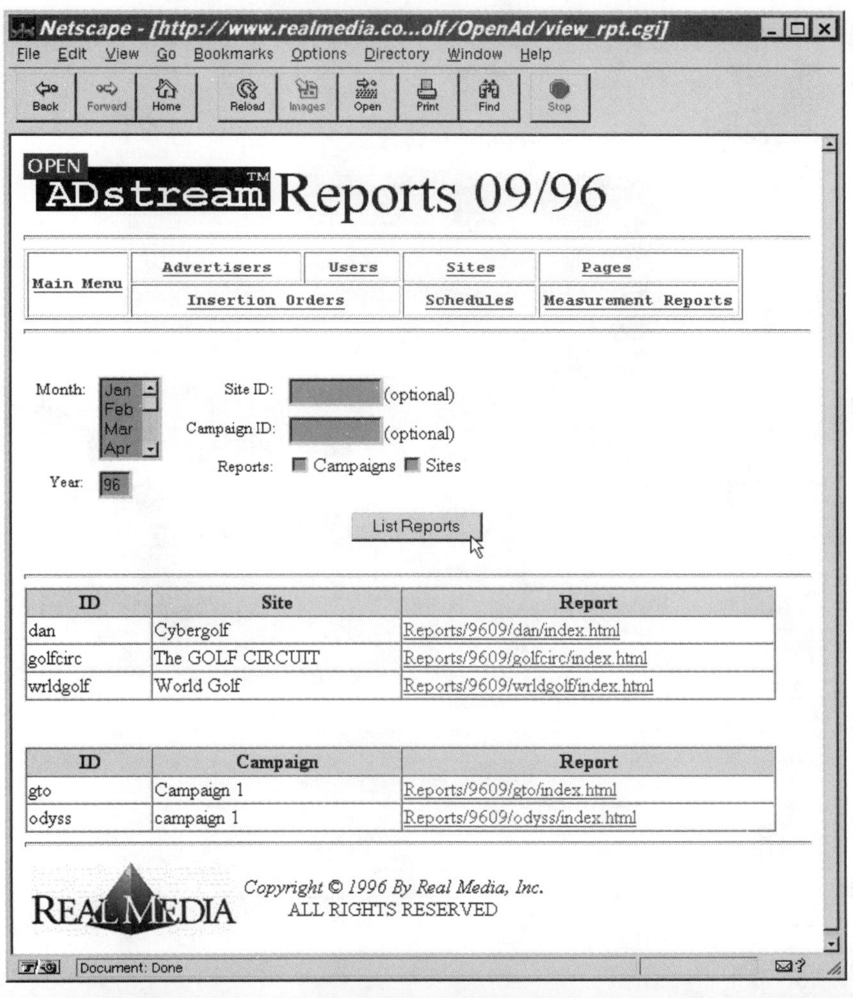

Figure 15-8. *The Reports page shows the available reports for each site and campaign*

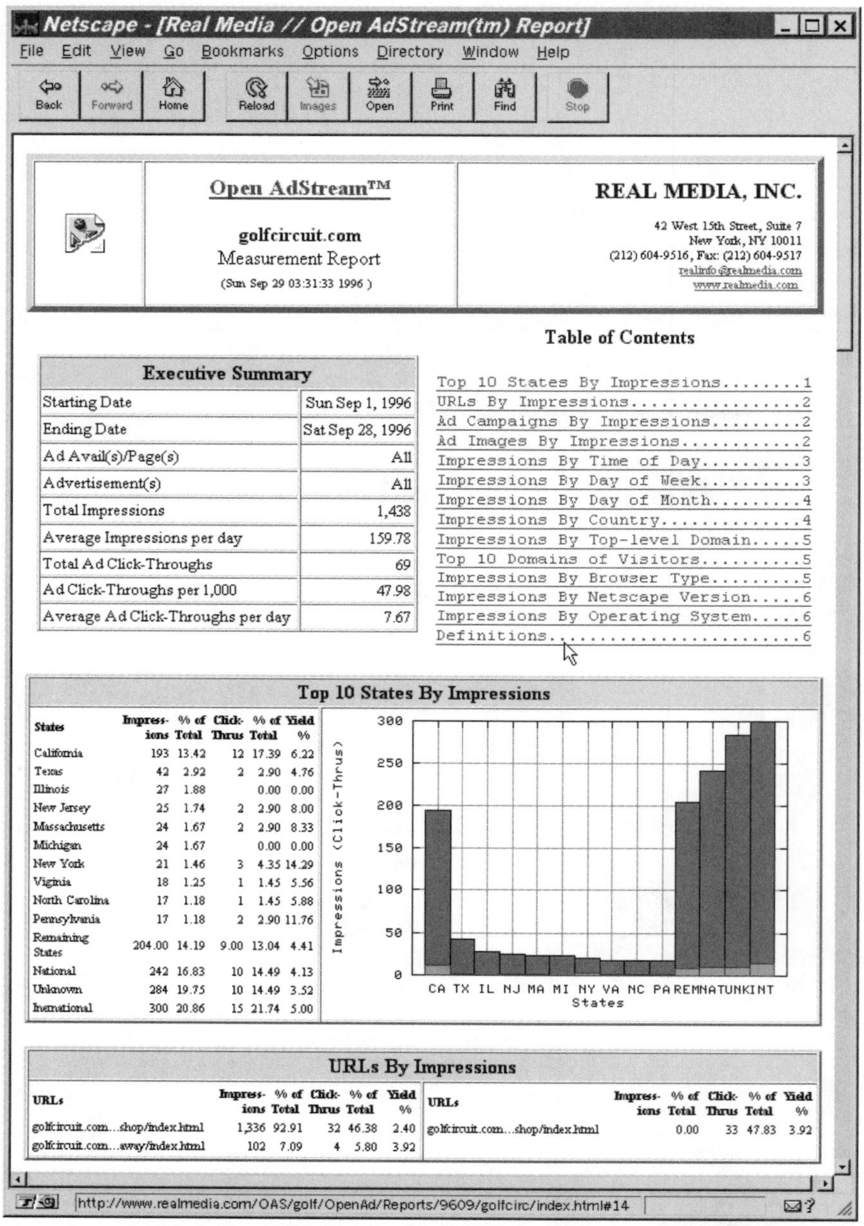

Figure 15-9. *The Open AdStream site reports are similar in many ways to log analysis reports, but they focus exclusively on ad banner views and click-throughs*

Open AdStream's reports are clean and professional. In addition, you can print them out to send them to your advertisers, or you can give them their own account (through the user administration screens) to view your ad traffic reports directly.

Other Commercial Ad Server Products

You can write your own simple ad server software or use Real Media's software just for letting them place ads on your site; in addition, other companies are vying for your software budget for ad server solutions. One such company is NetGravity; their product is called, simply, AdServer. NetGravity bills AdServer as the premiere solution for managing advertising on an Internet web site. You can use AdServer for banner placement, rotation, tracking, targeting, and reporting.

AdServer is Unix-based, with versions for Sun, SGI, DEC, HP, IBM, and Linux. I can't tell you a great deal more about this product, because NetGravity wasn't forthcoming with the details. They do offer an online, live demo on their web site at www.netgravity.com. Unfortunately, you can't take the tour immediately. You've got to fill out a questionnaire, after which the company will call you back—if interested. I can tell you that AdServer has a hefty price tag; at $20,000, it's a serious budget entry. For more details, see the web site at netgravity.com.

Another hot ad server product is ClickWise from ClickOver of Palo Alto, California. Like NetGravity's AdServer product, ClickWise seems to offer all of the features you would want from a centralized ad server, and it's quite a bit cheaper than AdServer at $6,000 per CPU. If you'd like more information on ClickWise, check out ClickOver's web site at www.clickover.com.

In the next chapter, we'll look at ad tracking and auditing services that are independent of the functions of managing campaigns and serving ads.

COLOR
FILM
AUDIO

Chapter Sixteen

Ad Tracking and Auditing Services

In Chapter 15, you learned about ad networks and ad servers. You probably realized as you read that chapter that the line between ad networks and ad servers is sometimes fuzzy—especially given that some ad networks either use their own centralized software or they provide quality ad server software to their members to use free of charge. However, just as there are third-party service companies that analyze server log files, there are also third-party firms that analyze ad activity.

In this chapter, we will take a minute to consider what the products (or services) of these companies are. Then, we will look at two of the predominant service companies in this field and see how they get the information that they need, and the types of reports that they provide.

Overview of Ad Tracking Services

Western International Media, Inc. is the largest wholesale buyer of advertising media in the world. Western brokers advertising spots ranging from blocks of television and radio time to billboards along the highways of North America and everything in between—including bulk spots on the World Wide Web. In June of 1996, Western announced in a press release that they would buy no more web ad space unless the space came with a guarantee of third-party audits and opinions as to the actual level of activity of those spots.

As a huge buyer of media, Western probably sets more trends than it follows, and we will probably see more web advertisers and wholesalers taking the same route in the future. This is the only outcome that makes any sense. Whenever any new industry becomes truly serious and the money gets big, the people putting up the money tend to demand accountability, which is what's happening with web advertising. It's not that the web has experienced rampant fraud or misrepresentations; this is just the way the business has to develop.

This brings up the question: "Just what is the product of a third-party ad tracking audit?" The product of a financial audit is clearly defined in the form of a letter from an auditing firm expressing a serious opinion about the financial condition of a company or organization. This letter is backed up by hundreds—or even thousands—of hours of statistical sampling, testing, and verification.

The product of audits of web-based ads and ad campaigns aren't as nearly well defined. For some advertisers and media buyers, third-party verification may be as simple as being able to view generated traffic statistic reports on the web pages of a tracking service like that of NetCount or Focalink. Others (like

Western) want more. They're going to be asking for, and getting, the very reputation of tracking service companies staked on the reasonable accuracy of the traffic reports. They'll get it in writing and signed in ink.

Several of the ad servers we looked at in Chapter 15 technically do much, if not all, of the same reporting that ad audit companies do. If you generate the reports yourself, however, you can't say they were generated by a third party.

NetCount's AdCount

NetCount is the leading server traffic analysis service company. We looked at NetCount's server traffic analysis service at length in Chapter 10. AdCount is another of NetCount's major service offerings, tailored specifically to producing vital statistics on ad activity. AdCount provides the following reports:

- Secured online reports
- Daily and weekly totals for ad impressions and click-throughs
- Daily and weekly totals for ad transfers
- Advertiser detail ad and summary reports
- Publisher detail ad and summary reports

NetCount's AdCount reports are secured on NetCount's web pages with basic user authentication—the same way their traffic statistics reports are. So only you, and anyone you give the username and password to, can see the reports. If the reason you subscribe to AdCount is to provide that all-important third-party verification to an advertiser, you'll need to provide the advertiser with the username and password so they can see the reports.

AdCount's Ad Impressions and Click-Through Report (shown in Figure 16-1) presents a table with weekly comparative numbers for impressions, click-throughs, and click-through rates. Following this table is a tabular and graphical representation of impressions and click-throughs on a daily basis for the week.

The Click-Throughs and Ad Transfers Report is similar to the Ad Impressions and Click-Through Report, but it adds data for ad transfers showing the actual number of successful transfers and the percentages of those successful transfers to total ad impressions and click-throughs. You can see a sample of this report in Figure 16-2; you can see that the percentage of

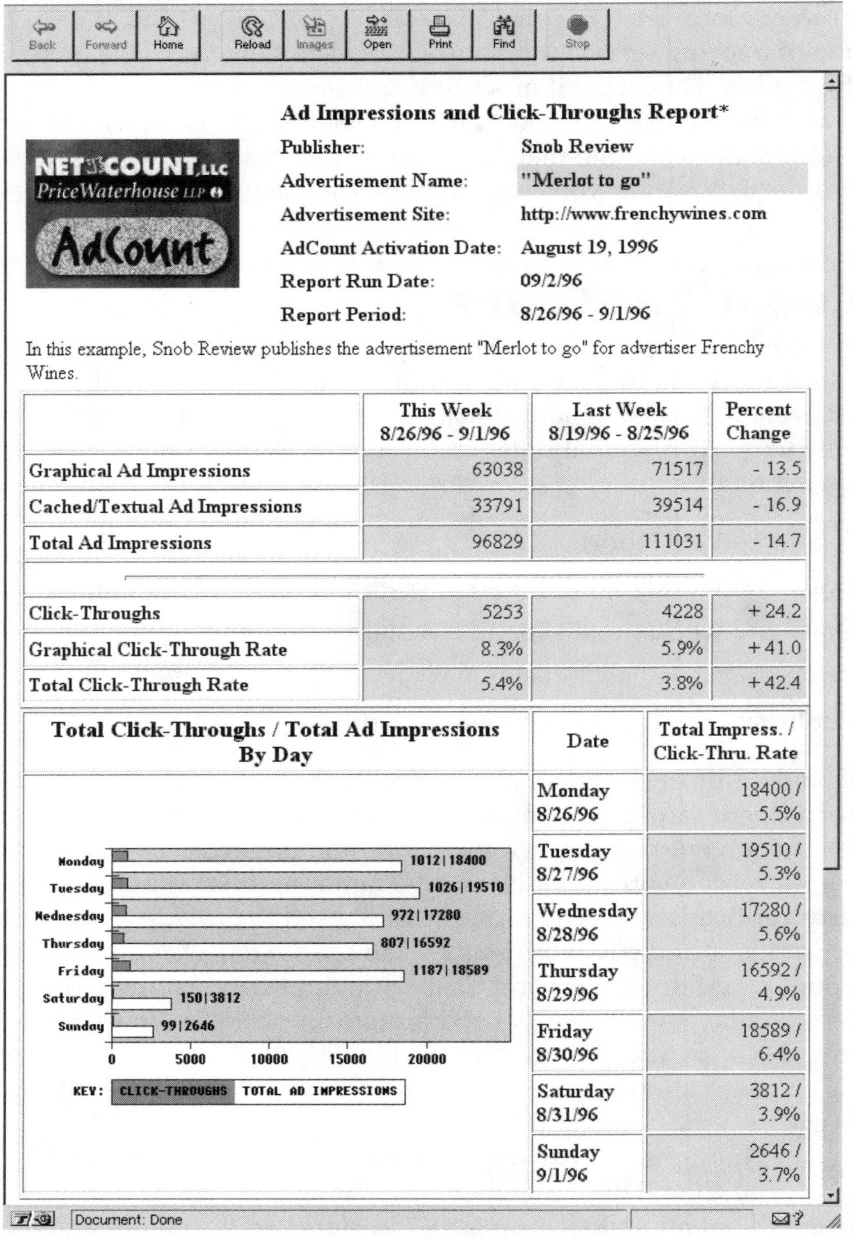
Ad Impressions and Click-Throughs Report*

Publisher:	Snob Review
Advertisement Name:	"Merlot to go"
Advertisement Site:	http://www.frenchywines.com
AdCount Activation Date:	August 19, 1996
Report Run Date:	09/2/96
Report Period:	8/26/96 - 9/1/96

In this example, Snob Review publishes the advertisement "Merlot to go" for advertiser Frenchy Wines.

	This Week 8/26/96 - 9/1/96	Last Week 8/19/96 - 8/25/96	Percent Change
Graphical Ad Impressions	63038	71517	- 13.5
Cached/Textual Ad Impressions	33791	39514	- 16.9
Total Ad Impressions	96829	111031	- 14.7
Click-Throughs	5253	4228	+ 24.2
Graphical Click-Through Rate	8.3%	5.9%	+ 41.0
Total Click-Through Rate	5.4%	3.8%	+ 42.4

Date	Total Impress. / Click-Thru. Rate
Monday 8/26/96	18400 / 5.5%
Tuesday 8/27/96	19510 / 5.3%
Wednesday 8/28/96	17280 / 5.6%
Thursday 8/29/96	16592 / 4.9%
Friday 8/30/96	18589 / 6.4%
Saturday 8/31/96	3812 / 3.9%
Sunday 9/1/96	2646 / 3.7%

Figure 16-1. *AdCount's Ad Impressions and Click-Through Report shows weekly comparisons for impressions, click-throughs, and click-through rates*

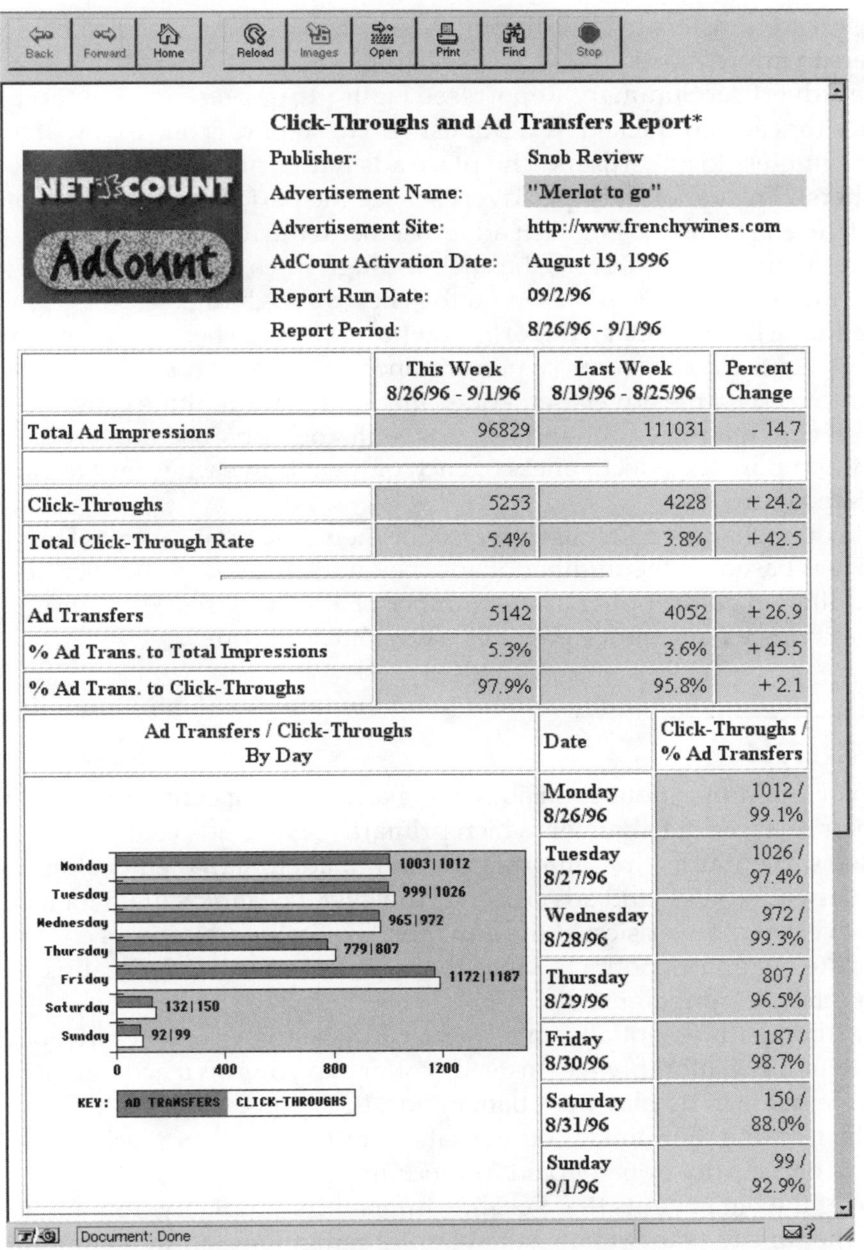

Back Forward Home Reload Images Open Print Find Stop

Click-Throughs and Ad Transfers Report*

NET COUNT AdCount

Publisher:	Snob Review
Advertisement Name:	"Merlot to go"
Advertisement Site:	http://www.frenchywines.com
AdCount Activation Date:	August 19, 1996
Report Run Date:	09/2/96
Report Period:	8/26/96 - 9/1/96

	This Week 8/26/96 - 9/1/96	Last Week 8/19/96 - 8/25/96	Percent Change
Total Ad Impressions	96829	111031	- 14.7
Click-Throughs	5253	4228	+ 24.2
Total Click-Through Rate	5.4%	3.8%	+ 42.5
Ad Transfers	5142	4052	+ 26.9
% Ad Trans. to Total Impressions	5.3%	3.6%	+ 45.5
% Ad Trans. to Click-Throughs	97.9%	95.8%	+ 2.1

Ad Transfers / Click-Throughs By Day

Date	Click-Throughs / % Ad Transfers
Monday 8/26/96	1012 / 99.1%
Tuesday 8/27/96	1026 / 97.4%
Wednesday 8/28/96	972 / 99.3%
Thursday 8/29/96	807 / 96.5%
Friday 8/30/96	1187 / 98.7%
Saturday 8/31/96	150 / 88.0%
Sunday 9/1/96	99 / 92.9%

Document: Done

Figure 16-2. *AdCount's Click-Throughs and Ad Transfers Report*

successful ad transfers to click-throughs is quite high while the percentage of transfers to impressions is low.

The Advertiser Summary Report (see Figure 16-3) presents comparable statistics for each ad tracked by a publisher. This report is one that will be of primary interest to advertisers who place ads with a number of sites (or publishers). This report lets the advertiser see the performance of each of its ads with each publisher. For each advertisement, the report shows the number of impressions, click-throughs, the click-through rates, ad transfers, and a relative ranking for each ad based on the click-through rate.

The Publisher Summary Report is a report for publishers (or sites). This report presents the comparative performances of each ad on a site ordered and subtotaled by the advertiser. For example, if you run three different ads for a golf club manufacturer and two ads for a golf apparel company, you can directly compare the effectiveness of each of the ads in each group to each other (see Figure 16-4).

NetCount charges a flat monthly fee for their AdCount service. The fee structure is based on the number of ads they track for you, with the rate for each additional ad dropping as the number of ads increases. For more information, see NetCount's pages at www.netcount.com.

Focalink

Focalink Communications is an ad serving and tracking company. Focalink calls their service SmartBanner, which primarily serves ads centrally out of their two server farms (in California and in Philadelphia); their customers are ad networks, sites, and advertisers. However, Focalink's main focus is the advertiser/buyer side of the equation. Typically, advertisers contact Focalink to negotiate contracts. Once the contract is finalized, Focalink tracks the advertiser's campaigns.

SmartBanner is essentially an ad server, but customers don't buy Focalink's software or pay a monthly fee. Instead, you make your own arrangements with advertisers to display their banner ads, then you contract with Focalink to do the tracking. Focalink's compensation for their tracking services is a small cut of the price you get paid for your impressions.

One of the big benefits that Focalink offers is its simplicity. You don't have to take the time to plan and manage your ad campaigns. Focalink does this for you. The service the company sells is in handling the details. The software runs on Focalink's own computers, and they handle setup, monitoring the ad serving, and the reporting. Just about the only thing that Focalink does not do

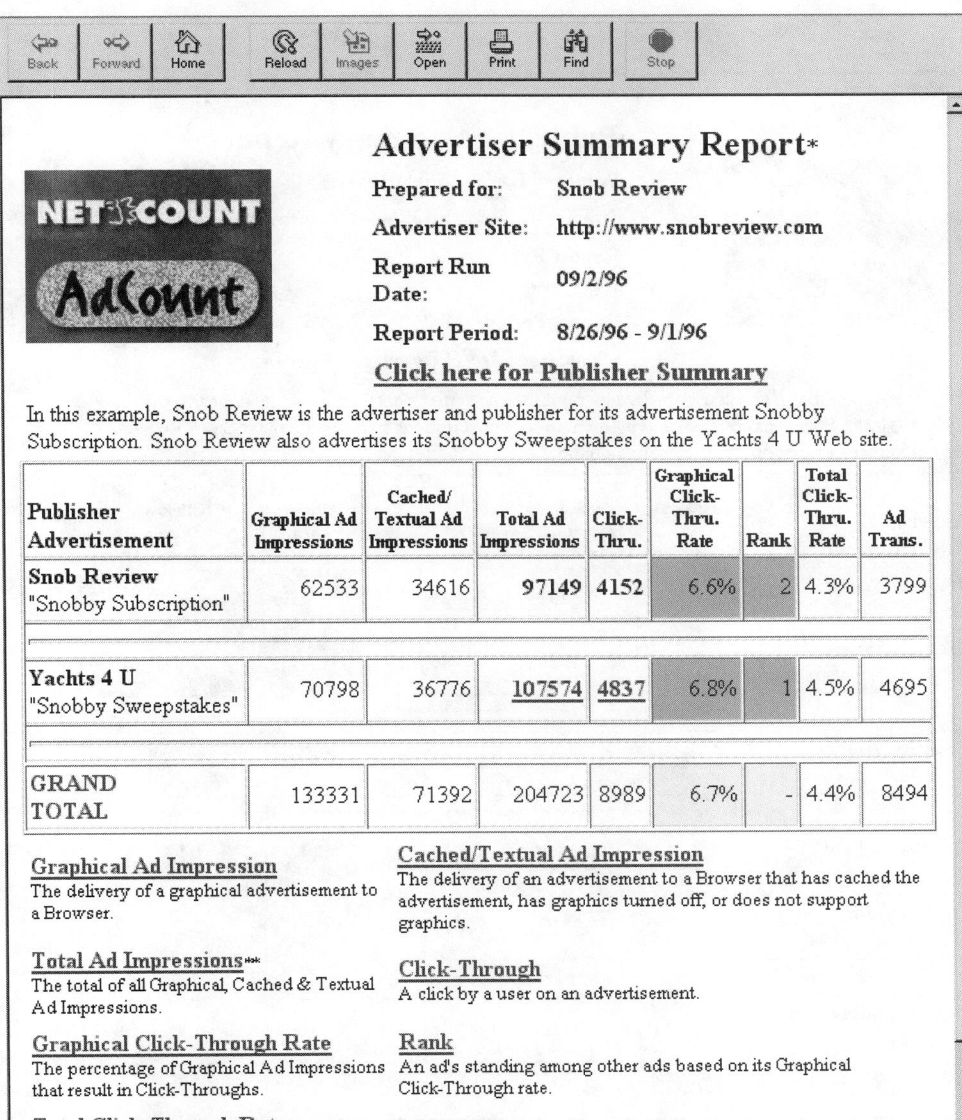
Back
Forward
Home
Reload
Images
Open
Print
Find
Stop
NET COUNT
AdCount
Advertiser Summary Report*
Prepared for: Snob Review
Advertiser Site: http://www.snobreview.com
Report Run Date: 09/2/96
Report Period: 8/26/96 - 9/1/96
Click here for Publisher Summary
In this example, Snob Review is the advertiser and publisher for its advertisement Snobby Subscription. Snob Review also advertises its Snobby Sweepstakes on the Yachts 4 U Web site.

Publisher Advertisement	Graphical Ad Impressions	Cached/ Textual Ad Impressions	Total Ad Impressions	Click-Thru.	Graphical Click-Thru. Rate	Rank	Total Click-Thru. Rate	Ad Trans.
Snob Review "Snobby Subscription"	62533	34616	97149	4152	6.6%	2	4.3%	3799
Yachts 4 U "Snobby Sweepstakes"	70798	36776	107574	4837	6.8%	1	4.5%	4695
GRAND TOTAL	133331	71392	204723	8989	6.7%	-	4.4%	8494

Graphical Ad Impression
The delivery of a graphical advertisement to a Browser.
Cached/Textual Ad Impression
The delivery of an advertisement to a Browser that has cached the advertisement, has graphics turned off, or does not support graphics.
Total Ad Impressions**
The total of all Graphical, Cached & Textual Ad Impressions.
Click-Through
A click by a user on an advertisement.
Graphical Click-Through Rate
The percentage of Graphical Ad Impressions that result in Click-Throughs.
Rank
An ad's standing among other ads based on its Graphical Click-Through rate.
Total Click-Through Rate
The percentage of Total Ad Impressions that result in Click-Throughs.
Ad Transfer
The successful arrival of a user at the advertiser's Web site.
Document: Done

Figure 16-3. *The Advertiser Summary Report compares ads for each site*

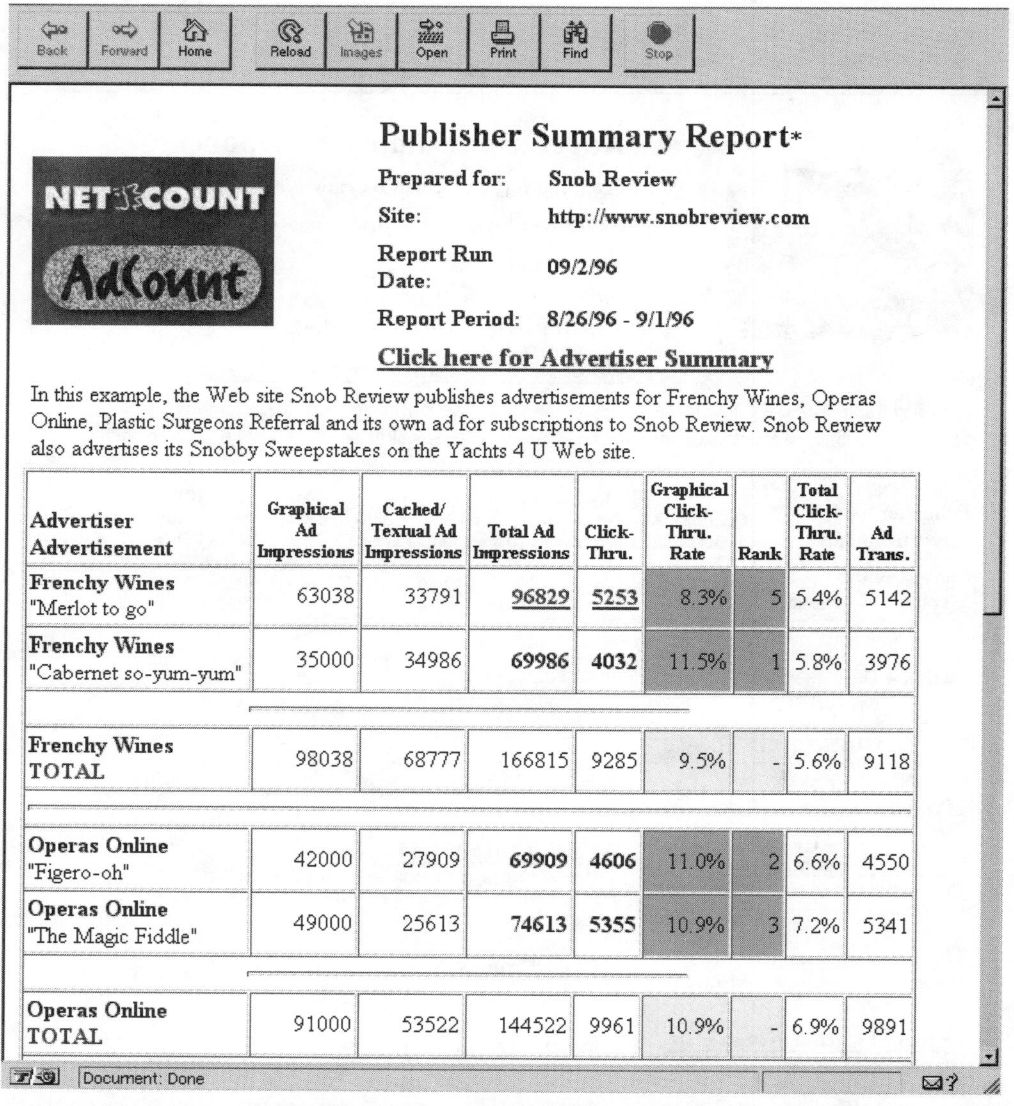

Publisher Summary Report*

Prepared for: Snob Review

Site: http://www.snobreview.com

Report Run Date: 09/2/96

Report Period: 8/26/96 - 9/1/96

Click here for Advertiser Summary

In this example, the Web site Snob Review publishes advertisements for Frenchy Wines, Operas Online, Plastic Surgeons Referral and its own ad for subscriptions to Snob Review. Snob Review also advertises its Snobby Sweepstakes on the Yachts 4 U Web site.

Advertiser Advertisement	Graphical Ad Impressions	Cached/ Textual Ad Impressions	Total Ad Impressions	Click-Thru.	Graphical Click-Thru. Rate	Rank	Total Click-Thru. Rate	Ad Trans.
Frenchy Wines "Merlot to go"	63038	33791	**96829**	**5253**	8.3%	5	5.4%	5142
Frenchy Wines "Cabernet so-yum-yum"	35000	34986	**69986**	**4032**	11.5%	1	5.8%	3976
Frenchy Wines TOTAL	98038	68777	166815	9285	9.5%	-	5.6%	9118
Operas Online "Figero-oh"	42000	27909	**69909**	**4606**	11.0%	2	6.6%	4550
Operas Online "The Magic Fiddle"	49000	25613	**74613**	**5355**	10.9%	3	7.2%	5341
Operas Online TOTAL	91000	53522	144522	9961	10.9%	-	6.9%	9891

Figure 16-4. *The Publisher Summary Report compares ads for each advertiser on a site*

is get involved in the creative aspects of an ad campaign: They won't do the graphics, and they won't make the decisions as to which ads should run on which sites or how often. This is between you and your advertisers.

Another benefit is that if you run a campaign across multiple sites, Focalink can report on the performances of each, showing the relative effectiveness of ads and campaigns by each individual site. Because Focalink handles the serving, tracking, and reporting centrally, you can be sure of consistent reporting for every site—in other words, you're truly comparing apples and apples.

By the time you read this, Focalink will also be able to distribute parts of their software to contracting sites. This will allow advertisers to aggregate a campaign strategy and reporting across multiple sites. Rather than reporting on each site individually, reports can be combined and comparatives can be presented.

We need to make one distinction here. While Focalink is an ad tracking company, they are not an auditing firm. The reports that they run on their systems are generated by their own software, and they don't produce audit reports. However, they are still considered a neutral third party which produces reports from their own (relatively) independent sources: their own software, which physically serves the ads.

Another service that Focalink offers is called MarketMatch. MarketMatch is a database registry where sites wanting to sell ad space can go to offer their pages to advertisers. The advertisers use this database to search for sites that closely fit their targeting desires. Anyone can enter the MarketMatch database, but whether or not the site will be attractive to advertisers will depend on the type of site, the volume of traffic, and other variables.

For more information about Focalink, visit their web pages at www.focalink.com. In the next chapter, we will look at two final topics that relate to marketing and advertising on the web—user registration systems and serving content dynamically.

Chapter Seventeen

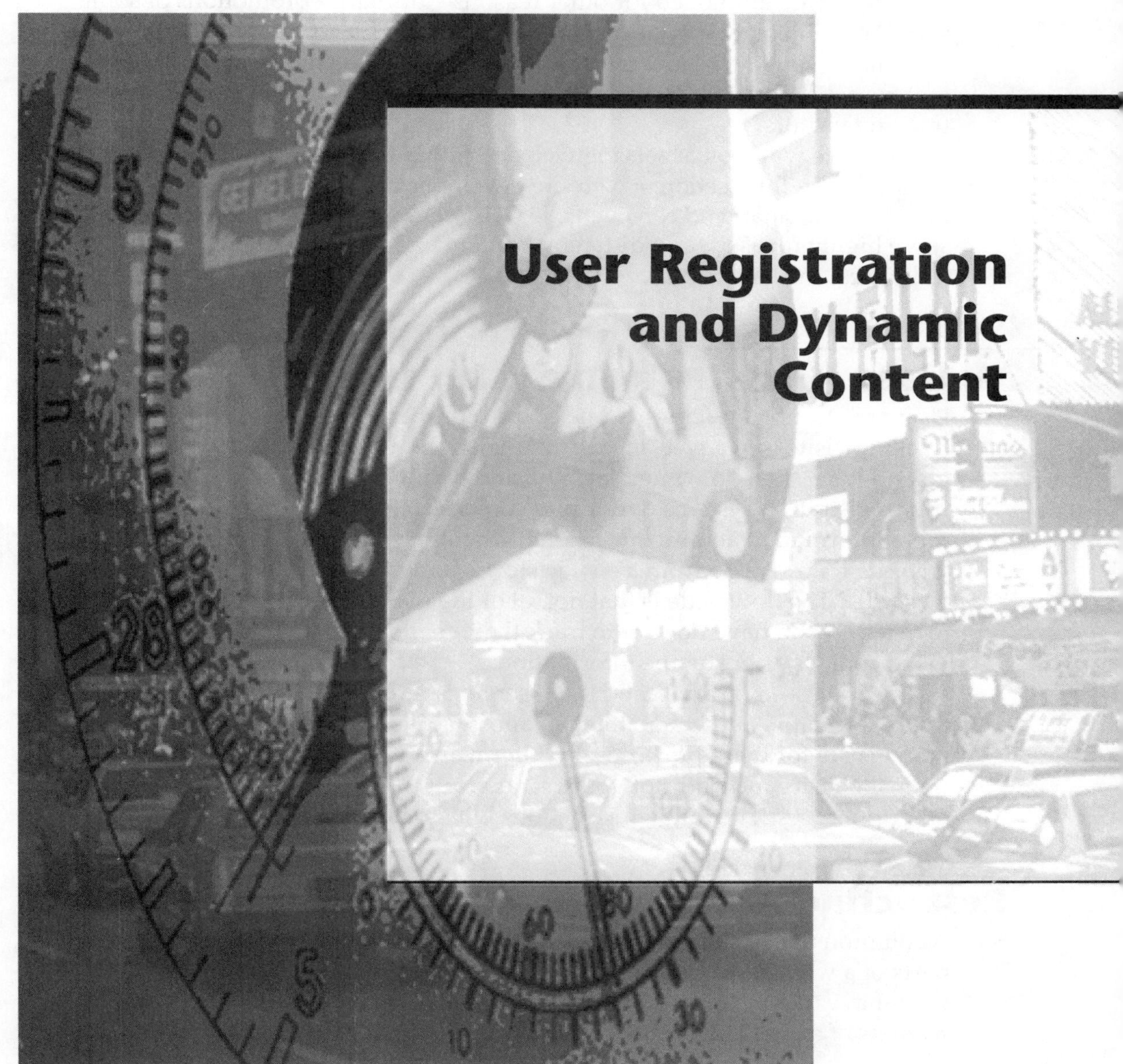

User Registration and Dynamic Content

In Chapter 7, we looked in detail at tracking visitors and visits by using HTTP cookies. In this chapter, we will look at two topics that impact both visit tracking and advertising on the web: user registration systems and serving site content dynamically.

Most often, user registration systems are not employed strictly for tracking visitors but are employed for other reasons—primarily promotions of some sort. We will look briefly at each of these applications of user registration systems so you know how registration systems are being used, then we will look very briefly at using them with a focus on reader demographics and tracking visitors.

Then we will look at serving web site content dynamically. Using the techniques in this section, a web site developer can track reader actions with a high level of detail and can even use CGI program and database integration to serve highly focused (and different) content and advertisements to each individual reader.

User Registration Systems

User registration systems are one of the oldest and most reliable methods for tracking visitors to a web site. Using registration, you can track where people go and what they do while they're visiting your site. The main advantage to user registration systems over using cookies is that you can get a great deal more information about your readers when they fill out a form and "join" your web site than you do from a cookie storing only HTTP header information or a user ID. The downside is that not all of the readers visiting your site will register, so you have no way to track these readers other than falling back on using cookies or hostname information.

Several common applications of user registration systems go beyond merely tracking visitors and where they go; most implementations of user registration systems are for these purposes. In other words, user registration can be used to track the visits and paths of your readers, but they are more often used for other purposes. The following sections will go over these applications.

Restricting Access

A common use of a user registration system is to restrict access to particular parts of a web site to registered users only. Most often, with such a system, users have to register and possibly pay a fee to access the "members only" sections of a site. These sites were prevalent when the web was first taking off, then they were used less, as webmasters and site owners realized that people prefer to go where they can get information and goodies anonymously and for free. However, plenty of sites on the web still provide content that is private or that has enough inherent value to warrant monthly or periodic charges.

For example, the web site of a county medical society might have public content plus a special section for physicians only—the society could use this section to inform member physicians of new treatments and current news topics. Such a medical society would probably want to provide these benefits only to doctors in the county who pay their dues to the medical society. In order to ensure that only member physicians have access, the society can ask doctors to register on the site; after the society validates their status, it can grant access to the registered doctors by issuing them a username and password. The society can then use basic user authentication to validate member physicians and restrict access to the members-only areas of the site.

In this example, it's clear that the primary reason for using registration is to restrict access. It's an added bonus that web servers can log the authuser name entered by members, enabling log-analysis software to track user movements through the site.

Additional Value Content

For another example, take the ESPN web site. As shown in Figure 17-1, ESPN offers a wealth of sports-related information and resources for all readers visiting the site. ESPN also offers content that has particular value above and beyond the scores of the latest baseball game. In Figure 17-1, you can see several buttons on the button bar and links in the content with a small image of a ticket stub next to them. To access these resources, the user must be registered on the ESPN site. If readers click on these links and aren't registered, they are encouraged to do so at that time.

Most of the time, ESPN charges a small monthly fee to become a registered user of the site; often, however, they run promotions that allow the user to register for a period of time free of charge. When I shot this figure of the ESPN site, it was running the following promotion: if the user downloaded Microsoft's Internet Explorer browser, he or she would be registered on the ESPN site free of charge for a period of time. More than likely, Microsoft paid ESPN the equivalent of the user registration fee, because of the good chance that the user would actually try out Internet Explorer if it were downloaded (and, in fact, when the user went to log in to any members-only areas, he or she was notified that Internet Explorer had to be used).

Figure 17-1. *ESPN's web site requires user registration for much of its most interesting and informative content*

In this example, ESPN is clearly using user registration because they can make money by selling the premium content on their web site. In the narrow view, you could say that ESPN does it for the same reason as the medical society in the earlier example—to restrict access to certain parts of the site for unregistered readers—but the bigger picture is clearly to make a little extra money. This bigger picture has nothing to do with tracking the browsing habits of readers, although, once again, from the time that registered readers log in to a restricted area, it is possible to track their movements through the site.

Contests

Probably the most common use of user registration today is to collect information on readers by offering promotions such as free giveaways and contests. Figure 17-2 shows the registration page for KFOG, a local radio station in the San Francisco area. Anyone who registers with the station gets a free newsletter and is entered into drawings for free CDs and tickets to shows. KFOG expressly states that they will not sell or trade the names of people who register on this site.

So why does KFOG offer user registration? According to their site manager, the goal is to develop a demographic database of their listeners in order to get to know them better and to facilitate communications with them through the web site. In addition, the web site promotes the radio station and the radio station promotes the web site—a synergistic combination of media.

Other web sites are not quite as ethical about their motives for collecting user registration data. The allure of the dollar can be overwhelming for many smaller web sites. With telemarketing firms offering between $.75 and several dollars (depending on the level of focus of the registrant) for names and demographic information, many small web site administrators find the bait irresistible.

Generally, web sites that offer registration for promotions do so with high motives: to promote interest and enthusiasm about the site. They do it to attract visitors and to lure visitors to return for repeat visits. Even most site owners that *do* sell registered user information to telemarketing and mail-order firms and others do so honestly—at least by omission—by not saying that they won't sell your name and address information.

Through these examples, we have identified four beneficial reasons to offer user registration to readers:

- To restrict access to private or confidential content
- To enhance site revenues by selling premium content

- To promote the site to encourage return visits
- To enhance site revenues by selling mailing or telephone lists

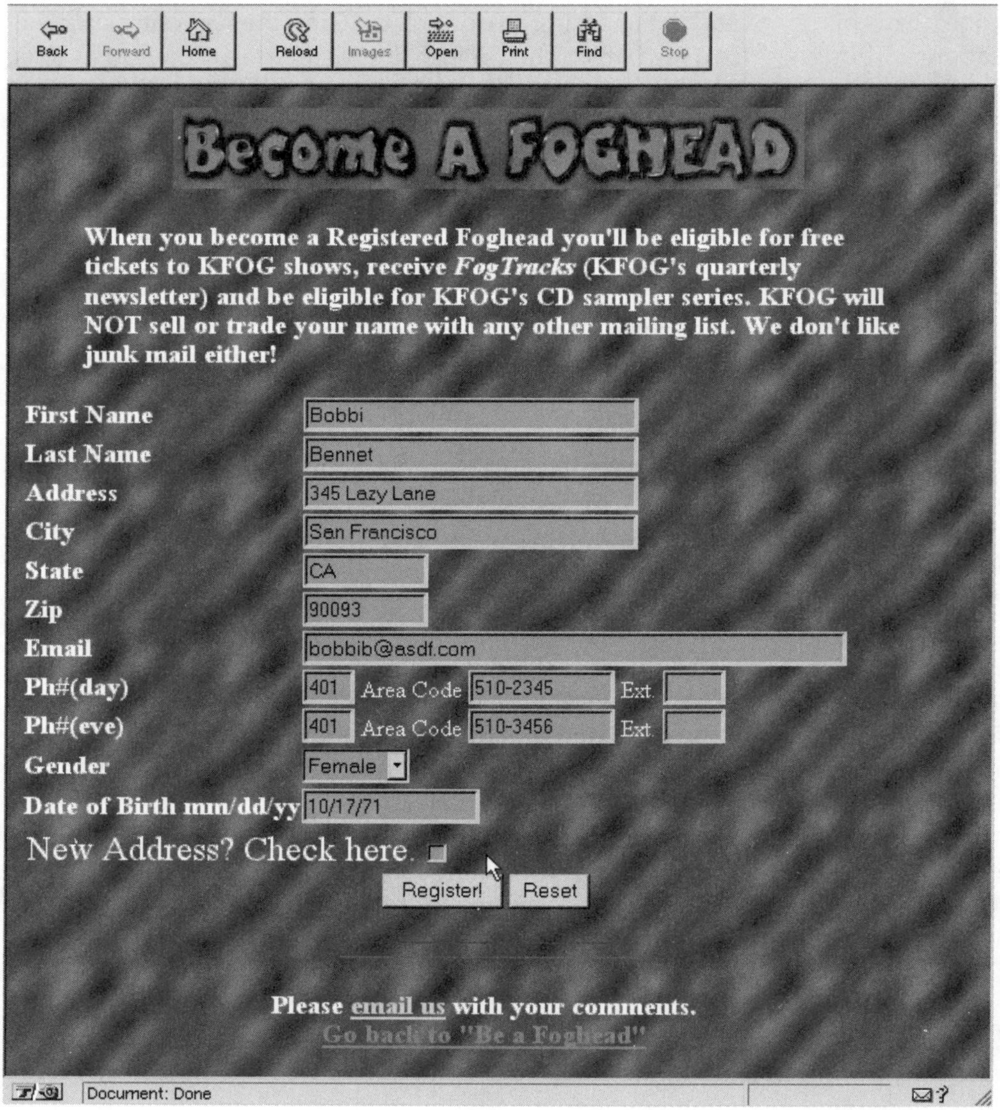

Back Forward Home Reload Images Open Print Find Stop

Become A FOGHEAD

When you become a Registered Foghead you'll be eligible for free tickets to KFOG shows, receive *FogTracks* (KFOG's quarterly newsletter) and be eligible for KFOG's CD sampler series. KFOG will NOT sell or trade your name with any other mailing list. We don't like junk mail either!

First Name: Bobbi
Last Name: Bennet
Address: 345 Lazy Lane
City: San Francisco
State: CA
Zip: 90093
Email: bobbib@asdf.com
Ph#(day): 401 Area Code 510-2345 Ext.
Ph#(eve): 401 Area Code 510-3456 Ext.
Gender: Female
Date of Birth mm/dd/yy: 10/17/71
New Address? Check here.

Register! Reset

Please email us with your comments.
Go back to "Be a Foghead"

Document: Done

Figure 17-2. *The user registration page for radio station KFOG*

In each of the three examples we looked at, using user registration information to track the paths and actions of visitors is a secondary, or even a minor, motivation. It's actually quite rare that a site owner will choose to implement a user-registration system primarily to track visitors. However, it is possible, and it does happen.

So how do you track visits by using basic user authentication? The easiest way is to use log-analysis software that takes authenticated user information into account (recall from Chapter 9 that Intersé's Market Focus package is one of the packages that currently does just that). Remember also (from Chapter 2) that web servers log the username of authorized users in the authuser field of the transfer log. After logging in, that username is logged on every subsequent hit generated by the reader. If Market Focus finds an authenticated username in the transfer log, that's the handle it needs to identify that visitor.

Using CGI to Serve Content Dynamically

The vast majority of web sites on the World Wide Web today are powered by one of a handful of the most popular standard web servers. The most popular of these are the free ones: the NCSA's httpd and the Apache server. However, commercial servers are quickly gaining popularity and support. The most prevalent of the commercial servers are those developed by Netscape (the Enterprise and FastTrack servers) and O'Reilly and Associates's WebSite.

In addition, the vast majority of sites powered by these common web servers serve static HTML documents. That is, most of the pages in sites are actual files containing HTML code sitting in the server's document directory or a subdirectory in the document directory hierarchy. For example, when you request the top-level page from a web server (say, http://www.asdf.com), the server will look for a file named index.html in the document directory for this site, and index.html is a real file that the server sends back to the requesting browser as a stream of characters.

Every page that you might view on a web site, however, isn't necessarily a rendering of a file sitting on the server computer. For example, when you use a search facility at a site to find something you're interested in, the page that you view showing you the results of your search would almost never be the contents of a file on the server. Instead, the page is generated on the fly by the program that performs the search—a CGI program.

This situation is a common one, because many sites deploy search tools to assist readers in finding what they want. Although a search tool is a very small component of most web sites, it is an excellent example of serving content dynamically. The page on which you view the results of a search is generated by the program and passed through the web server rather than being served directly by the server.

In Chapter 7, we explored the anatomy of HTTP requests and responses and we looked at how servers or CGI programs can create and send HTTP headers to set cookies. In Chapter 14, we looked at using CGI programs to redirect readers' browsers to other pages or other web sites. Using CGI programs with these two techniques is at the heart of the technology of serving content dynamically. Beyond these concepts, all you need to be able to serve content dynamically is a preferred scripting or programming language.

Here's a simple example. Let's say that a reader enters the URL http://www.asdf.com/cgi-bin/apage.pl into his or her web browser. Let's also say that apage.pl is a Perl script sitting in the cgi-bin directory of this site, and it contains the code in the following listing:

```
#!/usr/local/bin/perl
# APAGE.PL - Generates an HTML page

print <<EOT;
Content-type: text/html\n\n
<html><head><title>Page Title</title></head>
<body><center>
<h3>This page was generated by a CGI script.</h3>
</center></body></html>
EOT
```

The lines between the print statement and the EOT label essentially write the text between them to the browser. The first line of this text is an HTTP response header that notifies the browser that what's following is a stream of characters to be interpreted as HTML code. The browser can't make any distinction between HTML that it receives this way (generated by a program) or HTML that comes from a file served by the web server.

For most of you reading this book, this is pretty basic stuff; however, others may have less technical backgrounds and may appreciate the groundwork. The key here is that since HTML can be generated by a program, it can contain custom links handmade for each visitor. This way, when readers click on a hypertext link on a page, they notify the CGI program generating pages of not only the content they want to see (by which link they click on) but also of their own unique identity.

Let's do a new CGI script based on the apage.pl script above. We'll call this script dynapage.pl. Let's first modify the script to take an argument. If the script is called with no argument (as in the first time the script runs), it will serve a default or top-level page. The two hypertext links coded into the page include parameters. The link to go to a News page includes the argument NEWS, and the link to the Weather page includes the argument WEATHER.

```
#!/usr/local/bin/perl
# DYNAPAGE.PL - Serves three HTML pages dynamically

if ($ARGV[0]) {
  if ($ARGV[0] eq 'NEWS') {
    newspage();
  }
  if ($ARGV[0] eq 'WEATHER') {
    weatherpage();
  }
} else {
  defaultpage();
}

sub defaultpage {
print <<EOP;
Content-type: text/html\n\n
<html><head><title>Top Page</title></head>
<body><center>
<h3>This is the top-level page</h3>
<a href="/cgi-bin/dynapage.pl?NEWS">Click here for news</a><br>
<a href="/cgi-bin/dynapage.pl?WEATHER">Click here for weather</a>
</center></body></html>
EOP
}

sub newspage {
print <<EOP;
Content-type: text/html\n\n
<html><head><title>News</title></head>
<body><center>
<h3>The News Page</h3>
<a href="/cgi-bin/dynapage.pl?WEATHER">Click here for weather</a><br>
<a href="/cgi-bin/dynapage.pl">Top page</a>
</center></body></html>
EOP
}

sub weatherpage {
print <<EOP;
Content-type: text/html\n\n
<html><head><title>Weather</title></head>
<body><center>
<h3>The Weather Page</h3>
<a href="/cgi-bin/dynapage.pl?NEWS">Click here for news</a><br>
<a href="/cgi-bin/dynapage.pl">Top page</a>
</center></body></html>
EOP
}
```

When readers click on one of the two links on the top-level page, the same CGI program is called again, but this time it gets passed the relevant argument.

In other words, all three pages are generated by the same CGI script —dynamically.

This example is a highly simplistic example of serving content dynamically, but it serves to lay the foundation that you will need for where we're going with this. It's highly unlikely that anyone would develop an entire web site by hard-coding the HTML in a CGI program, but it is possible.

On the other hand, think of the possibilities that open up when you know you can use CGI scripts this way. Instead of generating links with arguments like NEWS and WEATHER, you can generate links that include as arguments the paths to actual HTML files on the disk. This program would be much smaller than the script just shown, plus it would be general in nature. It would just check to make sure that the referenced HTML file exists, then open it and send it to the reader's browser.

Your first reaction to that might be, "If real files exist on the disk, why not reference them directly and do away with all of this CGI nonsense?" That is a good question. If you really were only serving plain vanilla HTML files that reside in the server's document tree, you wouldn't go through all of this. But think in terms of combining the power of the CGI program with the flexibility of static HTML files.

Using these techniques, when readers first enter your site, you can have the CGI program issue them a unique identification number and code that number into every hypertext link on the page as an extra argument (the first being the path to the document to which the link refers). This way, every time readers click on a link in a document, their browsers send the path and name of the file they are requesting and their unique identification numbers.

Taking this to the next step is easy. Because you are in effect serving an entire web site dynamically through a single, small, generalized CGI program, making the program log each request—and in as much detail as you like—is as easy as opening a log file and writing the data to it. Voilà—absolute and concrete visit tracking.

This is the theory and a few of the basics of one way to serve content dynamically. Altogether, the prospects are very exciting. Some web technology companies have already taken this well beyond the prospective stage. The premier example is Webthreads of Vienna, Virginia. Webthreads has taken these basic concepts and woven a highly sophisticated system for serving web content dynamically and tracking visitors and their actions in real time.

In fact, Webthreads goes far beyond what we've described here. The Webthreads core technology is tightly integrated with a database management system and programmatic intelligence that will choose what content to present to readers. For example, an onscreen table of contents can show not only where you are in a web site but where you've been. It can even make suggestions as to where you might go next based on what you've been looking at and what you haven't seen yet.

You can probably guess just how powerful this technology could become, especially when it's combined with other techniques you've learned about in this chapter and book. Integrated with user registration for demographics and the techniques we covered in Chapters 14 and 15 for serving advertisement impressions intelligently, such a system could offer the ultimate in focused content and targeted advertising. For more information on Webthreads, see the accompanying CD or their web site at http://www.webthreads.com.

COLOR
FILM

PART FIVE

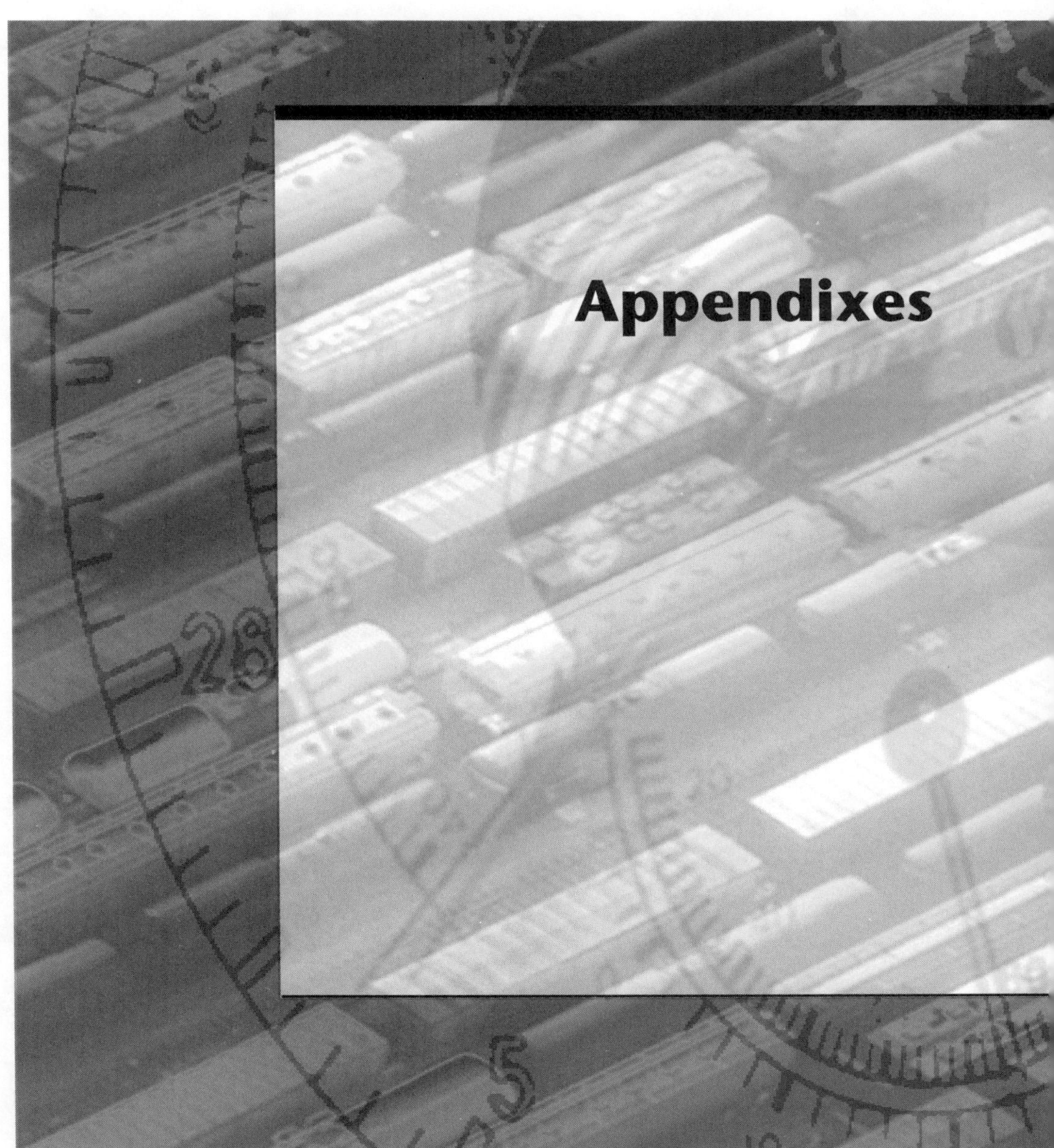

Appendixes

Appendix A

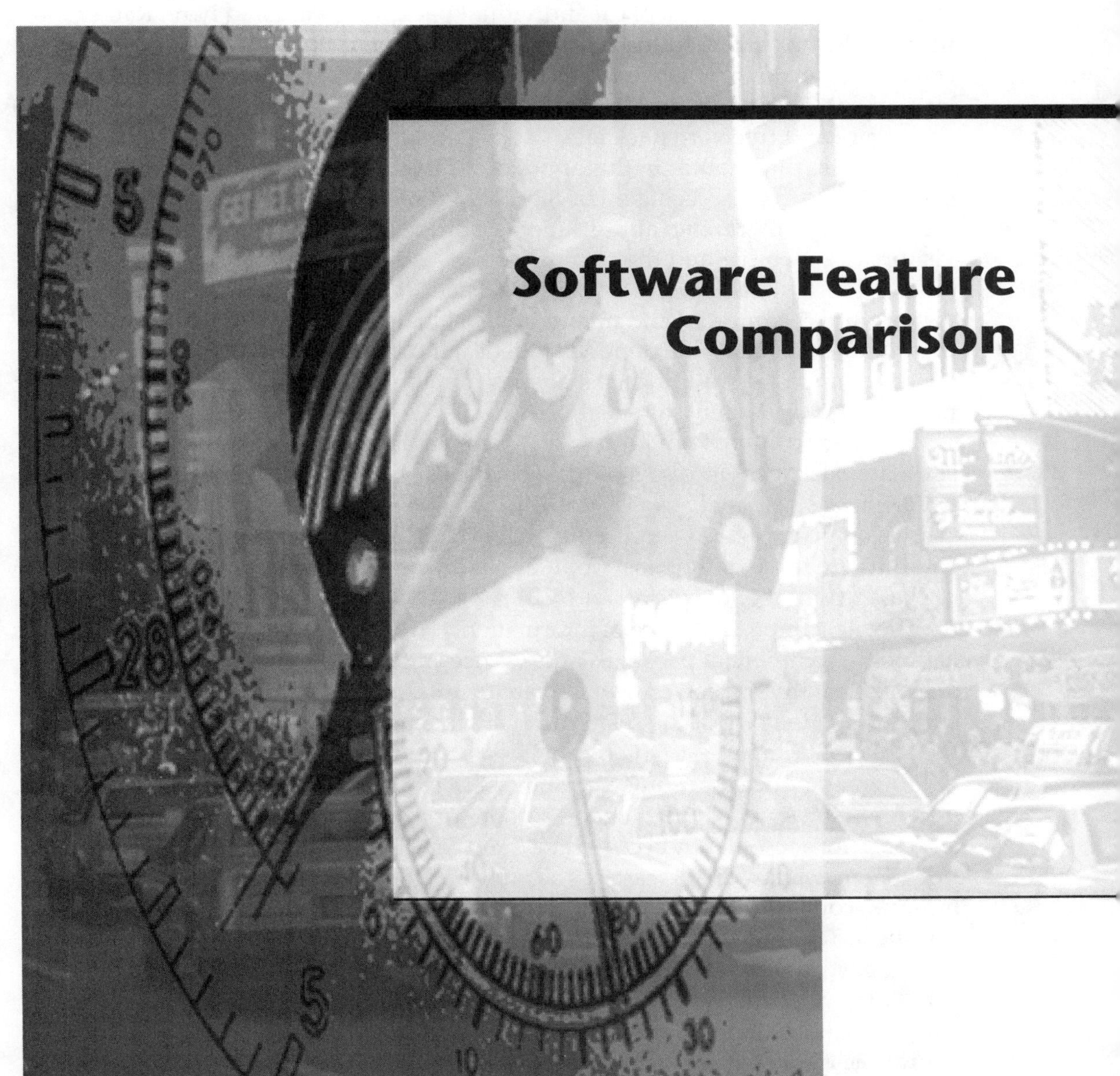

Software Feature Comparison

The tables in this appendix present a comparative analysis of the features offered by several of the most popular commercial log analysis software packages, NetCount (the most popular log analysis service), and several of the freeware/shareware log analysis tools. By no means do these tables represent a comprehensive analysis of all of the log analysis tools available nor even exhaustive analyses of the individual packages evaluated here. Rather, these tables are intended to give you an idea of the number and types of statistics that different log analysis tools may offer and to compare the general level of complexity of commercial packages and services to that of the shareware and freeware packages.

There are three tables in this appendix. Table A-1 (request-focus statistics) lists statistics that can be taken or tabulated directly from a server log file; these statistics require no algorithm to distinguish between the number of visitors to a web site and the number of requests in the transfer log. Table A-2 (visit-focused statistics) shows those packages that do calculate a number of visitors and the number and nature of the statistics that are derived based on that number. Table A-3 (extended statistics) indicates the extent to which these packages go to infer additional statistics from data, such as hostnames, referrer, and user agent information.

The three columns on the left side of the tables contain data for commercial software packages (Intersé's Market Focus, e.g. Software's WebTrends, and net.Genesis's net.Analysis Desktop). The next column is the one log analysis service company that is represented here (NetCount). The three right-hand columns are three of the most popular shareware or freeware software packages (wusage, wwwstat, and VBStats).

The most important piece of information you should get from this appendix is that the commercial packages and services offer much more depth in their analysis than their freeware/shareware counterparts. That should be apparent simply from looking at the relative density of check marks on the left side of the tables as compared to the right.

You may notice that some statistics seem to be redundant or repeated in different tables, and there's a good reason for it. For example, Table A-1 shows a statistic for "Top sites by access count," and Table A-2 has a statistic for "Visits from organizations." In this case, it's actually bad to have a check mark in "Top sites by access count" in Table A-1, yet it's good to have one in the corresponding statistic in Table A-2. "Top sites by access count" represents the number of hits from a particular host or domain name. In the accounting world, this is something referred to as a hash total—it's a meaningless number. Unknown variables like the number of (and which) pages were viewed and the number of graphic images and other inline objects on those pages make this statistic

Statistic	Intersé Market Focus	e.g. Software WebTrends	net.Genesis net.Analysis Desktop	NetCount	wusage	wwwstat	VBStats
Number of requests	✓	✓	✓	✓	✓	✓	✓
Number/percentage of successful or failed requests		✓		✓			
Number/percentage of cached requests		✓					
Client errors (number of requests and percentage)		✓	✓	✓			
Server errors (number of requests and percentage)		✓	✓				
Top documents or files (most requested documents)	✓	✓	✓	✓	✓		✓
Document transfers (number) by day				✓			
Top downloaded files by type (all files)		✓	✓	✓			
Top submitted forms and scripts		✓					
Bottom documents or files	✓						
Top documents by directory	✓						
Top directories accessed		✓					
Average number of requests per week	✓			✓			

Table A-1. *Request-Focused Statistics*

Statistic	Intersé Market Focus	e.g. Software WebTrends	net.Genesis net.Analysis Desktop	NetCount	wusage	wwwstat	VBStats
Average number of requests per day	✓	✓		✓			
Total bytes transferred	✓		✓		✓	✓	✓
Average bytes transferred by day	✓						
Average bytes transferred by hour of the day	✓				✓		
Average number of hits on weekdays		✓					
Average number of hits on weekends		✓					
Most/least active day of the week (and number of hits)		✓					
Most/least active day ever (number of hits)		✓					
Activity level by day of week/hour of day		✓					
Activity by organization type (hits, percentage)		✓					
Top sites by access count					✓		✓
Top sites by byte count							✓
Top domains by access count					✓		
Accesses by result code					✓		

Table A-1. *Request-Focused Statistics* (continued)

Statistic	Intersé Market Focus	e.g. Software WebTrends	net.Genesis net.Analysis Desktop	NetCount	wusage	wwwstat	VBStats
Number of visits	✓	✓	✓	✓			
Average number of requests per visit	✓						
Average duration of a visit	✓						
Average number of visits per week	✓						
Average number of visits per day	✓	✓					
Visits (number) by hour of the day				✓			
Visits from organizations (most active organizations)	✓	✓	✓	✓			
Visits by organization type (root domain)	✓		✓	✓			
Visits from countries (most active countries)	✓	✓					
Top visit entry pages	✓						
Top page durations	✓						
Top exit pages (last pages)	✓						
Average number of users on weekdays/weekends		✓					
Visit level by day of week/hour of day		✓					

Table A-2. *Visit-Focused Statistics*

Statistic	Intersé Market Focus	e.g. Software WebTrends	net.Genesis net.Analysis Desktop	NetCount	wusage	wwwstat	VBStats
Top U.S. geographic regions	✓	✓					
Percentage of visits from inside/outside the U.S.		✓					
Top cities	✓	✓					
Top referring organizations	✓						
Top referring URLs	✓		✓				
Top browsers	✓		✓				
Top user operating systems	✓						

Table A-3. *Extended Statistics*

misleading at best. A single visit by a single visitor from a single organization may leave a track of a single hit from that organization, or the visitor may just as easily generate a couple hundred hits during the visit. If the software can't distinguish a visit from a hit, it can't possibly report a meaningful statistic regarding visits from organizations. You'll notice that the only packages that try to pass off statistics like these as meaningful are the shareware/freeware packages.

Appendix B

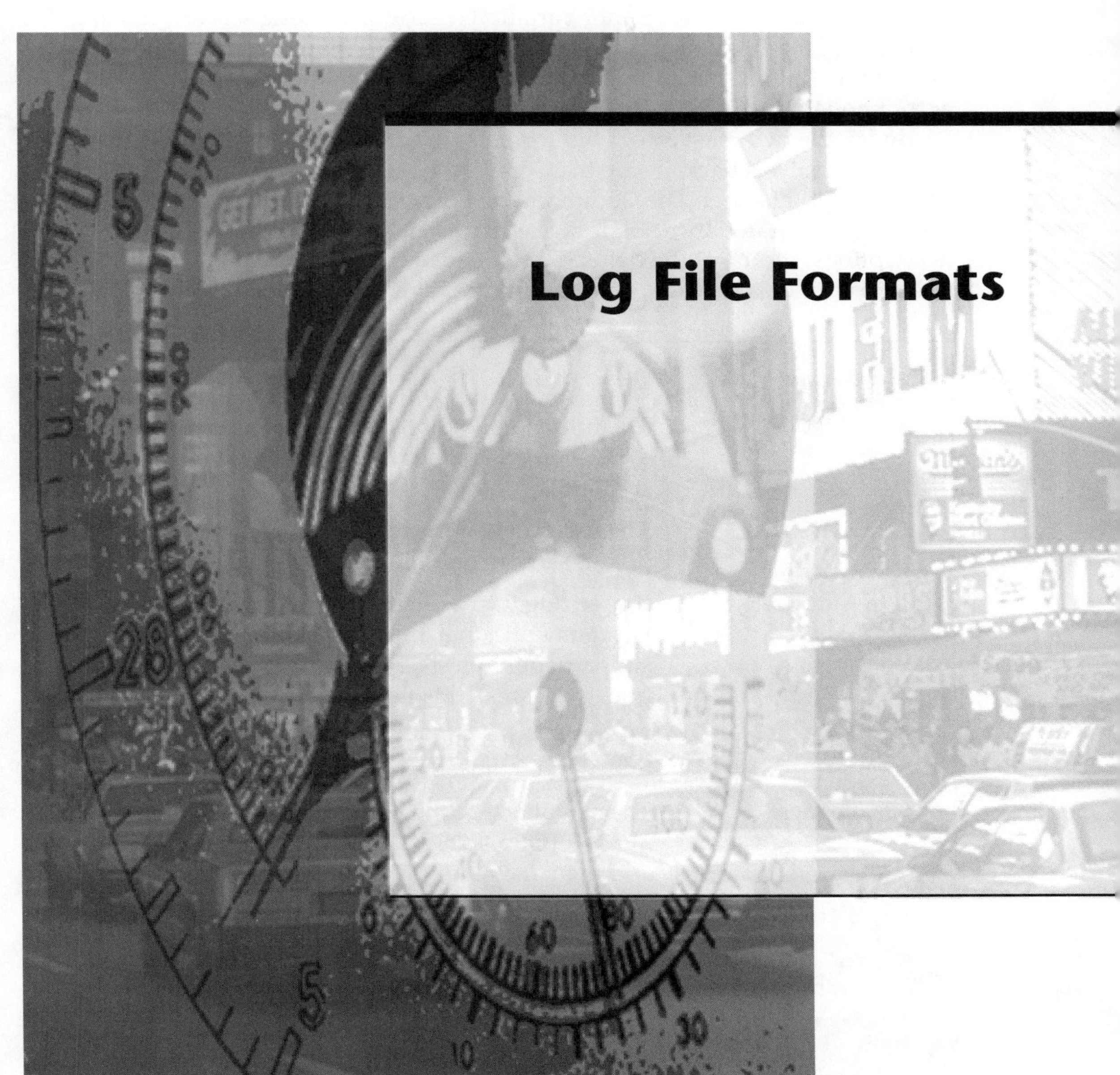

Log File Formats

Although the vast majority of web servers on the Internet are powered by one of only a half-dozen or so of the most popular server packages, literally dozens more are available. Unfortunately, the developers of some of these packages haven't seen the wisdom or the benefits of conforming to what has become fairly solidly accepted standards for log file formats. This has resulted in numerous minor variations in format among servers, making things difficult for analysis software developers. To their credit, some developers (Intersé, for example), have made valiant attempts to accommodate many of the more arcane server formats.

By far, the two most important formats that are directly usable by nearly every log analysis package are the Common Log Format and the extended (or combined) log format. The Common Log Format consists of seven fields of data, each separated by a single space character. Table B-1 presents short descriptions of each of the fields. The extended log format adds two additional fields to each log entry: the referrer and user agent fields. See Table B-2 for a

Field	Name	Description
1	Host	The fully qualified hostname or IP address of the system on which the requesting client (browser) is running.
2	RFC931	The username of the reader making the request. This field is seldom anything other than a hyphen (-).
3	Authuser	If users log into a secure area of a web site requiring user authentication, their username is recorded here on subsequent requests. For unauthenticated accesses (most accesses), this field contains only a hyphen (-).
4	Time stamp	The date, time, and GMT offset of the server at the time of the request.
5	Request	The actual text of the HTTP request. This starts with a method (GET, POST, and so on) and includes the path and filename requested.
6	Status	The three-digit status code of the server upon servicing the request.
7	Transfer volume	The number of bytes sent as a result of the request.

Table B-1. *Common Log Format Fields*

description of the extended log format fields. For more detailed information, see Chapters 2 through 5.

A third additional field—the cookie field—is quickly becoming common, although its inclusion isn't yet widespread enough to warrant a new format name. As more sites and webmasters realize the benefit of setting cookies to track visitors, more log analysis software will use the cookie field and more server developers will make including cookie information in their standard log files easy.

Field	Name	Description
1	Host	The fully qualified hostname or IP address of the system on which the requesting client (browser) is running.
2	RFC931	The username of the reader making the request. This field is seldom anything other than a hyphen (-).
3	Authuser	If users log into a secure area of a web site requiring user authentication, their username is recorded here on subsequent requests. For unauthenticated accesses (most accesses), this field contains only a hyphen (-).
4	Time stamp	The date, time, and GMT offset of the server at the time of the request.
5	Request	The actual text of the HTTP request. This starts with a method (GET, POST, and so on) and includes the path and filename requested.
6	Status	The three-digit status code of the server upon servicing the request.
7	Transfer volume	The number of bytes sent as a result of the request.
8	Referrer	The address (URL) of the page from which the address was obtained (usually in a hypertext link).
9	User agent	The name and version of the user agent software (browser) and, often, the operating system of the computer it's running on.

Table B-2. *Extended Log Format Fields*

COLOR
FILM

Appendix C

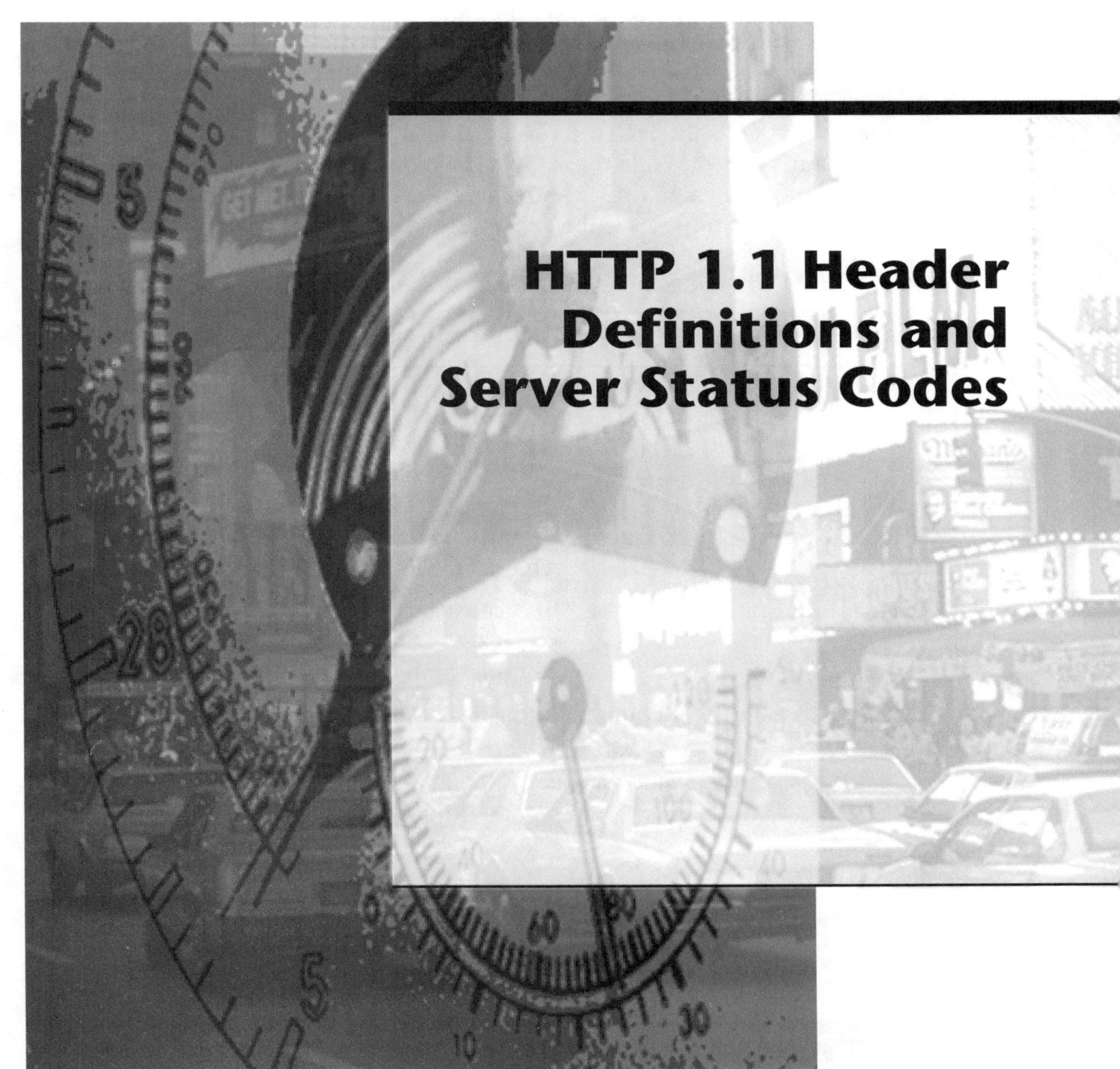

HTTP 1.1 Header Definitions and Server Status Codes

As this book was nearing completion, the World Wide Web Consortium (W3C) released the draft specification for version 1.1 of the HTTP protocol. Version 1.1 includes many new header types and status codes, and the potential for much more flexibility and functionality than the earlier 1.0 version. Although it will be many months—perhaps years—before most of the new functionality of version 1.1 is fully utilized in web servers, gateways, proxies, and browser software, I have decided to include in this appendix the complete set of HTTP 1.1 headers and status codes to give you a heads up on what's coming.

Table C-1 presents the HTTP 1.1 headers and Table C-2 presents the version 1.1 server status codes. While these tables are complete, they describe only the intended purpose of each header and status code. For more detailed information and examples, see the current draft specification at the W3C web site at http://www.w3.org.

The HTTP 1.0 headers and status codes that are in common use today with version 1.0 of the protocol are presented in bold in the tables.

Header	Type	Definition
Accept	Request	Sent by a browser to a server to indicate preferred media types.
Accept-Charset	Request	Sent by a browser to indicate to the server character sets acceptable to the browser.
Accept-Encoding	Request	Indicates encoding schemes acceptable to the browser, for example, ZIP, gzip, compress files.
Accept-Language	Request	Indicates acceptable natural languages (i.e. English, French).
Accept-Ranges	Response	Servers may indicate to browsers the acceptability of range requests.
Age	Response	Conveys an estimate of the time since a request.
Allow	Entity	Indicates to a browser the methods that are allowed for a given resource (for example, GET, HEAD, PUT).

Table C-1. *HTTP Version 1.1 Headers*

Header	Type	Definition
Authorization	Request	Clients (browsers) can use this header to request authentication.
Cache-Control	General	Indicates to any caching entities (gateways, proxies, browsers) directives relating to the caching of the relevant document; for example, "no-cache" indicates that the document should not be cached at all and "max-age=86400" means that the document should be cached for only one day.
Connection	General	HTTP 1.1 supports persistent connections. This header can be used in the format "Connection: close" to emulate HTTP 1.0 (nonpersistent behavior).
Content-Base	Entity	Like the <BASE> tag in HTML itself, this header can be used at the HTTP level to indicate the base URI for resolving relative URLs within an entity.
Content-Encoding	Entity	Indicates the type of any encoding present in an entity, for example, ZIP or gzip files.
Content-Language	Entity	Indicates the natural language of the content (i.e. English, French).
Content-Length	Entity	Indicates the size of the entity (in a decimal number of octets).
Content-Location	Entity	Specifies the location of other parts of multipart entities.
Content-MD5	Entity	Indicates and conveys an MD5 digest—similar to a checksum of the entity body. This is good for detecting transmission errors.
Content-Range	Entity	Sent with a partial entity body to indicate to a client where it should be inserted into a full entity body.

Table C-1. *HTTP Version 1.1 Headers* (continued)

Header	Type	Definition
Content-Type	Entity	Indicates the media type of the entity body, for example, "text/html" or "image/gif".
Date	General	This header is most often sent by servers so that browsers can evaluate it against cached copies. Browsers may send it with PUT and POST requests.
ETag	Entity	Defines the entity tag for the associated entity. (See If-Match.)
Expires	Entity	Provides the date and time after which the response (or entity) should be considered stale.
From	Request	Should contain the Internet e-mail address of the person controlling the browser making the request.
Host	Request	Specifies the hostname and port of the resource being requested. This will be used to distinguish between multiple hosts attached to a common IP address.
If-Match	Request	Makes a request conditional based on matching an entity tag (see ETag).
If-Modified-Since	Request	Makes a request conditional.
If-None-Match	Request	Makes a request conditional based on not matching an entity tag (see ETag).
If-Range	Request	Makes a conditional request for an entity.
If-Unmodified-Since	Request	Makes a request conditional.
Last-Modified	Entity	Indicates the date and time that a server believes an entity was last modified.
Location	Response	Redirects a user agent to another URL.
Max-Forwards	Request	Specifies the maximum number of hops that can be followed with the TRACE method.

Table C-1. *HTTP Version 1.1 Headers (*continued*)*

Header	Type	Definition
Pragma	General	Conveys directives to gateways, proxies, and user agents. This header still exists in version 1.1, but it will be going away eventually.
Proxy-Authenticate	Response	Informs a user agent that proxy authentication is required.
Proxy-Authorization	Request	Carries proxy authentication information (username and password) to a proxy server.
Public	Response	Servers can use this header to inform user agents of the methods (GET, PUT, POST, and so on) supported by the server.
Range	General	Used to specify byte ranges in an entity; for example, when a partial entity is cached, a request could be made for only the unavailable part of the entity—if the entity is unchanged.
Referer	Request	Carries the URL of the resource from which the request address was obtained.
Retry-After	Response	This can be used to notify user agents of how long a service will be unavailable.
Server	Response	Carries the name and version of the server software.
Transfer-Encoding	General	Indicates the type (if any) of encoding done to the entity body for transmission.
Upgrade	General	Notifies the recipient of an on-the-fly protocol change, for example, a switch to a different HTTP version.
User-Agent	Request	Conveys the name, version number, and possibly operating system of a user agent to a server.

Table C-1. *HTTP Version 1.1 Headers* (continued)

Header	Type	Definition
Vary	Response	Indicates that a response entity was selected using server-driven negotiation.
Via	General	Used by gateways and proxies for holding the software name and version. This must be used instead of appending to the User-Agent header.
Warning	Response	Carries additional information about the status of response codes, if needed.
WWW-Authenticate	Response	Indicates that authorization is required for the requested resource, causing user agents to prompt for username and password.

Table C-1. *HTTP Version 1.1 Headers* (continued)

Status Code	Description
100	Continue
101	Switching Protocols
200	OK
201	Created
202	Accepted
203	Non-Authoritative Information
204	No Content
205	Reset Content
206	Partial Content
300	Multiple Choices

Table C-2. *HTTP Version 1.1 Server Status Codes*

Status Code	Description
301	Moved Permanently
302	Moved Temporarily
303	See Other
304	Not Modified
305	Use Proxy
400	Bad Request
401	Unauthorized
402	Payment Required
403	Forbidden
404	Not Found
405	Method Not Allowed
406	Not Acceptable
407	Proxy Authentication Required
408	Request Time-Out
409	Conflict
410	Gone
411	Length Required
412	Precondition Failed
413	Request Entity Too Large
414	Request-URI Too Large
415	Unsupported Media Type
500	Internal Server Error
501	Not Implemented
502	Bad Gateway
503	Service Unavailable

Table C-2. *HTTP Version 1.1 Server Status Codes* (continued)

Status Code	Description
504	Gateway Time-Out
505	HTTP Version Not Supported

Table C-2. *HTTP Version 1.1 Server Status Codes* (continued)

Appendix D

Log Analysis and Advertising-Related Companies

This appendix presents the names, mailing addresses, and web addresses of some of the companies involved in web server log analysis and Internet advertising. Represented here are most of the companies referred to in the book, and a few that are not. If you would like to add your company to this list for future editions, you can submit the information online at http://websitestats.com.

Bien Logic

2223 Avenida de la Playa
Suite 205
La Jolla, CA 92037
(619) 551-4888

http://www.bienlogic.com

Bien Logic is primarily a web site developer and Internet publishing business. However, Bien Logic also develops software, and the reason for inclusion here is their product SurfReport—a new, and somewhat unique log analysis package.

ClickOver

525 Alma Street
Palo Alto, CA 94301
(415) 322-8336

http://www.clickover.com

ClickOver is a relative newcomer with a very interesting ad management package called ClickWise. ClickWise is a graphical Java application with novel features, like buttons and slider bars, for controlling ad service.

DoubleClick

(888) 727-5300

http://www.doubleclick.com

DoubleClick is the largest of the advertising networks, and as such, probably offers the most highly targeted advertising on the Internet. You must be able to deliver a minimum of 100,000 impressions per month for consideration.

e.g. Software, Inc.

621 SW Morrison
Suite 1025
Portland, OR 97205
(503) 294-7025

http://www.egsoftware.com

e.g. Software is the developer of WebTrends, a commercial web server log analysis package. See Chapter 9 or the accompanying CD for more information.

Firewall Studios

649 Front Street
San Francisco, CA 94111
(415) 982-3344

http://www.firewallstudios.com

An Internet marketing company, Firewall develops custom programs for increasing site traffic by setting up contests and mailing lists for marketing companies to build user demographic databases.

Focalink Communications

459 Hamilton Avenue
Suite 306
Palo Alto, CA 94301
(415) 328-5465

http://www.focalink.com

Focalink serves ads for advertisers and for ad networks.

Forrester Research, Inc.

1033 Massachusetts Avenue
Cambridge, MA 02138
(617) 497-7090

http://www.forrester.com

Forrester is a contract research company focusing on technology and business. Forrester makes some of their excellent reports public on their web site. Check out their report on Internet advertising.

Group Cortex, Inc.

2300 Chestnut Street
Suite 230
Philadelphia, PA 19103
(215) 854-0646

http://www.cortex.net

Group Cortex is the developer of the SiteTrack log analysis software for Unix computers running any of the Netscape web servers. See Chapter 9 or the accompanying CD for more information.

IMGIS, Inc.

19200 Von Karman Avenue
Suite 690
Irvine, CA 92715
(800) 973-1174

http://www.imgis.com

Offers targeted advertising on their ad network and complete ad management with their service called AdForce.

Internet Link Exchange

http://www.linkexchange.com

Offers to serve ads in exchange for you serving ads of other network members on your site free of charge. They serve your ad once for every two impressions you provide.

Internet Profiles Corporation (I/PRO)

785 Market Street
13th Floor
San Francisco, CA 94103
(415) 975-5800

http://www.ipro.com

I/PRO is one of the two major log analysis service companies. I/PRO offers tracking, auditing, ad tracking, and centralized user registration services. See Chapter 10 or the accompanying CD for more information.

Intersé Corporation

111 W. Evelyn Avenue
Suite 213
Sunnyvale, CA 94086
(408) 732-0932

http://www.interse.com

Intersé is the developer of the Market Focus log analysis software package. See Chapter 9 or the accompanying CD for more information.

NarrowCast Media

(310) 979-4638

http://www.narrowcastmedia.com

Offers highly targeted ad space and guarantees a twenty percent click-through rate. NarrowCast has no minimum traffic requirements.

NetCount, LLC

1645 North Vine Street
Los Angeles, CA 90028
(213) 848-5700

http://www.netcount.com

NetCount is one of the two major log analysis service companies. See Chapter 10 or the accompanying CD for more information.

net.Genesis Corporation

68 Rogers Street
Cambridge, MA 02142-1119
(617) 577-9800

http://www.netgen.com

net.Genesis is the developer of the net.Analysis and net.Analysis Desktop log analysis packages. See Chapter 9 or the accompanying CD for more information.

NetGravity

1700 S. Amphlett Blvd.
Suite 350
San Mateo, CA 94402-2715
(415) 655-4777

http://www.netgravity.com

NetGravity is the developer of one of the leading ad server software packages. Called simply AdServer, NetGravity's server is also one of the most comprehensive and expensive ad servers available.

Open Market, Inc.

245 First Street
Cambridge, MA 02142
(617) 621-9500

535 Middlefield #250
Menlo Park, CA 94025
(415) 614-3400

http://www.openmarket.com

Open Market develops WebReporter, a log analysis software package. See Chapter 9 or the accompanying CD for more information.

Real Media

32 E. 31st Street
9th Floor
New York, NY 10016
(212) 725-4537

270 Commerce Drive
Suite 6000
Fort Washington, PA 19034
(215) 654-8376

http://www.realmedia.com

Real Media is an advertising services company (an ad network) that offers targeted advertising placement and detailed measurement across a large network of local sites. For joining their ad network, Real Media also offers its AdStream software for complex ad serving and reporting.

Streams Online Media Development

1313 North Wood Street
Chicago, IL 60622
(312) 342-7747

http://www.streams.com

Streams is the developer of Lilipad, a log analysis software system for Unix systems. See Chapter 9 or the accompanying CD for more information.

WebThreads

1919 Gallows Road
10th Floor
Vienna, VA 22182
(703) 848-9027

http://www.webthreads.com

WebThreads develops a product that makes serving and tracking an entire web site dynamically practical and easy. See Chapter 17 or the accompanying CD for more information.

World Wide Web Consortium (W3C)

http://www.w3.org

The W3C is the organization that hammers out and publishes the standards for the technologies involving the web—like the HTTP protocol, HTML, SGML, and so on. If you want the definitive answer on something to do with one of these topics, check the current draft specification here.

Index

NOTE: Page numbers in *italics* refer to illustrations or tables.

D

E

F

G

H

L

M

R

S

V

W

X

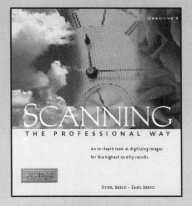

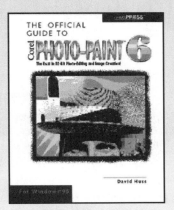

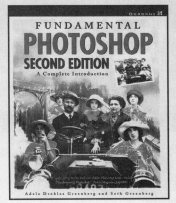

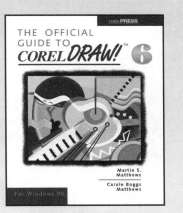

ORDER BOOKS DIRECTLY FROM OSBORNE/McGRAW-HILL

For a complete catalog of Osborne's books, call 510-549-6600 or write to us at 2600 Tenth Street, Berkeley, CA 94710

Call Toll-Free, *24 hours a day, 7 days a week, in the U.S.A.*
U.S.A.: 1-800-262-4729 ***Canada: 1-800-565-5758***

Mail ***in the U.S.A. to:***
McGraw-Hill, Inc.
Customer Service Dept.
P.O. Box 182607
Columbus, OH 43218-2607

Canada
McGraw-Hill Ryerson
Customer Service
300 Water Street
Whitby, Ontario L1N 9B6

Fax ***in the U.S.A. to:***
1-614-759-3644

Canada
1-800-463-5885
Canada
orders@mcgrawhill.ca

SHIP TO:

Name ____________________

Company ____________________

Address ____________________

City / State / Zip ____________________

Daytime Telephone *(We'll contact you if there's a question about your order.)*

ISBN #	BOOK TITLE	Quantity	Price	Total
0-07-88				
0-07-88				
0-07-88				
0-07-88				
0-07-88				
0-07088				
0-07-88				
0-07-88				
0-07-88				
0-07-88				
0-07-88				
0-07-88				
0-07-88				
0-07-88				
	Shipping & Handling Charge from Chart Below			
	Subtotal			
	Please Add Applicable State & Local Sales Tax			
	TOTAL			

Shipping & Handling Charges

Order Amount	U.S.	Outside U.S.
$15.00 - $24.99	$4.00	$6.00
$25.00 - $49.99	$5.00	$7.00
$50.00 - $74.99	$6.00	$8.00
$75.00 - and up	$7.00	$9.00
$100.00 - and up	$8.00	$10.00

Occasionally we allow other selected companies to use our mailing list. If you would prefer that we not include you in these extra mailings, please check here: ❑

METHOD OF PAYMENT

❑ Check or money order enclosed (payable to Osborne/McGraw-Hill)

❑ AMERICAN EXPRESS ❑ DISCOVER ❑ MasterCard ❑ VISA

Account No. ____________________

Expiration Date ____________________

Signature ____________________

In a hurry? Call with your order anytime, day or night, or visit your local bookstore.

Thank you for your order

Code BC640SL

About the CD

The CD attached to the back cover of this book contains evaluation copies of some of the best and most popular log analysis software on the market today. In addition to software, most of the contributing companies have also developed HTML presentations to show you how and why their particular software or service is the best for your needs. Two of the contributors, I/PRO and NetCount, LLC, are not software developers at all—these companies offer services to analyze your web server log files for you. Both I/PRO and NetCount have developed and contributed presentations to fill you in on the details of the services they offer. The contributors to the CD are the following:

- e.g. Software
- Group Cortex
- Intersé
- I/PRO
- NetCount, LLC
- net.Genesis
- Open Market
- Streams
- WebThreads

Using the CD is simple. On the root directory is a file named Start.htm. Just load this file into your web browser to begin. For example, if you use Netscape or Internet Explorer under Windows 95 and the drive letter of your CD-ROM drive is D, then just open the File menu, choose Open (or Open File) and type **D:\Start.htm** in the Open File dialog box. Or, of course, you can just open the drive window for your CD-ROM drive and double-click the Start.htm file or drag it onto a running web browser.

Once you can view the Start.htm file, read it for more detailed instructions on viewing the HTML presentations and installing any of the evaluation software onto your computer's hard drive. Follow the hypertext links you find by clicking on them with your mouse.

In some cases, the HTML presentations on the CD look very similar to the web sites of their respective companies. However, the content on this CD is actually custom-tailored by each company especially to the audience of this book. So if you've been to the company's web site and it looks similar, do take the time to go through its presentation here anyway.

Because with HTML there's no easy way to make a web browser fire off an executable program, you can't install the evaluation software directly through the web interface. There are, however, detailed instructions in the Start.htm document for installing each software package.

Happy exploring!

WARNING: BEFORE OPENING THE DISC PACKAGE, CAREFULLY READ THE TERMS AND CONDITIONS OF THE FOLLOWING COPYRIGHT STATEMENT AND LIMITED CD-ROM WARRANTY.

Copyright Statement

This software is protected by both United States copyright law and international copyright treaty provision. Except as noted in the contents of the CD-ROM, you must treat this software just like a book. However, you may copy it into a computer to be used and you may make archival copies of the software for the sole purpose of backing up the software and protecting your investment from loss. By saying, "just like a book," The McGraw-Hill Companies, Inc. ("Osborne/McGraw-Hill") means, for example, that this software may be used by any number of people and may be freely moved from one computer location to another, so long as there is no possibility of its being used at one location or on one computer while it is being used at another. Just as a book cannot be read by two different people in two different places at the same time, neither can the software be used by two different people in two different places at the same time.

Limited Warranty

Osborne/McGraw-Hill warrants the physical compact disc enclosed herein to be free of defects in materials and workmanship for a period of sixty days from the purchase date. If the CD included in your book has defects in materials or workmanship, please call McGraw-Hill at 1-800-217-0059, 9am to 5pm, Monday through Friday, Eastern Standard Time, and McGraw-Hill will replace the defective disc.

The entire and exclusive liability and remedy for breach of this Limited Warranty shall be limited to replacement of the defective disc, and shall not include or extend to any claim for or right to cover any other damages, including but not limited to, loss of profit, data, or use of the software, or special incidental, or consequential damages or other similar claims, even if Osborne/McGraw-Hill has been specifically advised of the possibility of such damages. In no event will Osborne/McGraw-Hill's liability for any damages to you or any other person ever exceed the lower of the suggested list price or actual price paid for the license to use the software, regardless of any form of the claim.

OSBORNE/McGRAW-HILL SPECIFICALLY DISCLAIMS ALL OTHER WARRANTIES, EXPRESS OR IMPLIED, INCLUDING BUT NOT LIMITED TO, ANY IMPLIED WARRANTY OF MERCHANTABILITY OR FITNESS FOR A PARTICULAR PURPOSE. Specifically, Osborne/McGraw-Hill make no representation or warranty that the software is fit for any particular purpose, and any implied warranty of merchantability is limited to the sixty-day duration of the Limited Warranty covering the physical disc only (and not the software), and is otherwise expressly and specifically disclaimed.

This limited warranty gives you specific legal rights; you may have others which may vary from state to state. Some states do not allow the exclusion of incidental or consequential damages, or the limitation on how long an implied warranty lasts, so some of the above may not apply to you.

This agreement constitutes the entire agreement between the parties relating to use of the Product. The terms of any purchase order shall have no effect on the terms of this Agreement. Failure of Osborne/McGraw-Hill to insist at any time on strict compliance with this Agreement shall not constitute a waiver of any rights under this Agreement. This Agreement shall be construed and governed in accordance with the laws of New York. If any provision of this Agreement is held to be contrary to law, that provision will be enforced to the maximum extent permissible, and the remaining provisions will remain in force and effect.

NO TECHNICAL SUPPORT IS PROVIDED WITH THIS CD-ROM.